DESIGNED FORESTS

Designed Forests: A Cultural History explores the unique kinship that exists between forests and spatial design; the forest's influence on architectural culture and practice; and the potentials and pitfalls of "forest thinking" for more sustainable and ethical ways of doing architecture today. It tackles these subjects by focusing on architecture's own dispositions, which stem from an ecology of metaphor that surrounds its encounters with the forest and undergird ideas about Nature and natural systems. The book weaves together global narratives and chapters explore a range of topics, such as the invention of forest plans in colonial India, the war waged on the jungles of Vietnam, economic land use concepts in rural Germany, precolonial ecological pasts in Manhattan, and technologically saturated forests in California. This book is essential for landscape architects, urbanists, architects, forestry experts, and everyone concerned with larger environmental contexts and the ever-evolving relationship between nature and culture.

Dan Handel is an architect and curator whose work focuses on underexplored ideas, figures, and practices that shape contemporary built environments. Over the past fifteen years, he has been studying the links between scientists, forest managers, and spatial designers, resulting in various exhibitions and publications on the subject.

"Forests cover over four billion hectares of our planet earth. In his brilliant book, Dan Handel reminds us that our forests project a powerful image of nature even though they are constructed, manmade and totally designed. Handel skillfully weaves together a riveting narrative inextricably linking forest to culture, geography, and our human condition."

Professor Brigitte Shim, *Daniels Faculty of Architecture, Landscape and Design, University of Toronto*

"In the process of asking questions about how forests are rationalized culturally, Handel advances the political control of reforestation, which ultimately reveals that making or conserving forests is a design project. Outside of the current environmental crisis, what this book offers is a richly illustrated account of the global movement to create, protect and plant forests, while provoking artists, designers and planners to consider their role in the affair."

Rosetta Elkin, *Academic Director, Master in Landscape Architecture Program, Pratt Institute*

"Handel's book establishes a parallel between two forms of cognition: 'thinking machines' and 'thinking forests.' This parallel has important implications for built and natural environments, particularly in relation to developments in artificial intelligence and discoveries about tree communication."

Phu Huang, *Head of Architecture, Knowlton School, Ohio State University*

DESIGNED FORESTS

A Cultural History

Dan Handel

Routledge
Taylor & Francis Group
NEW YORK AND LONDON

Cover image: Paolo Uccello, Hunt in the Forest, c. 1465. Painting in the collection of the Ashmolean Museum, University of Oxford.

First published 2025
by Routledge
605 Third Avenue, New York, NY 10158

and by Routledge
4 Park Square, Milton Park, Abingdon, Oxon, OX14 4RN

Routledge is an imprint of the Taylor & Francis Group, an informa business

© 2025 Dan Handel

Every effort has been made to contact copyright-holders. Please advise the publisher of any errors or omissions, and these will be corrected in subsequent editions.

ISBN: 978-1-032-75318-8 (hbk)
ISBN: 978-1-032-75317-1 (pbk)
ISBN: 978-1-003-47341-1 (ebk)

DOI: 10.4324/9781003473411

Typeset in Bembo
by SPi Technologies India Pvt Ltd (Straive)

CONTENTS

ABOUT THE AUTHOR

Dan Handel is an architect and curator. He develops research-based projects with special attention to underexplored ideas and practices that shape contemporary built environments. He curated various forest-related exhibitions, including *First, the Forests* at the Canadian Centre for Architecture (2012) and *Wood: The Cyclical Nature of Materials, Sites, and Ideas* at the Nieuwe Instituut in Rotterdam (2014). He has published and lectured extensively on the subject, and his writing has been included in the *Journal of Landscape Architecture* (spring 2011), *Cabinet Magazine* (winter 2012–2013), *The Word for World Is Still Forest* (K. Verlag, 2017), and *Harvard Design Magazine* (spring/summer 2018). He presented his work at the University of Toronto (2019), the Kunsthistorisches Institut in Florenz (2021), *Prada Frames* at the Salone del Mobile in Milan (2022), and the Zurich University of Applied Sciences (2023).

PREFACE

The road that led to the writing of this book was by no way a straight line. Beginning with a chance encounter with aerial photographs of the Checkboard Cascades in the Pacific Northwest circa 2009, I was fortunate to receive a traveling grant to go visit forestlands, production facilities, and Forest Service sites in Montana firsthand, which opened an entirely unfamiliar territory of possible connections between forestry and design. Initial intuitions turned into work on several exhibitions in research institutions, the writing of scholarly and non-scholarly texts, and the presentation of findings and ideas in various venues. This meandering path provided opportunities for many conversations without which this book could not have evolved and reached its current form.

I wish to thank the Penny White traveling grant at the Harvard Graduate School of Design for sending me on this trip; Dale Bosworth, former head of the US Forest Service, who generously hosted me in a house on a creek in Missoula and sketched out an illuminating history of designed forests in the region; and Felipe Correa, who helped navigate this project from its earliest inquiries.

As the project developed, I began to write down bits of it and send them for publication. I wish to extend my thanks to the individuals who accepted, edited, and published these texts, which served as blueprints for themes and figures that are featured in the current book: Kelly Shannon, who published the first scholarly product of this work in the *Journal of Landscape Architecture* and showed astonishing patience as I was cutting my teeth as a writer; Sina Najafi of Cabinet, who brilliantly commented and shaped a text on Dietrich Brandis, fueling my further investigations into the subject; Anna-Sophie Springer, who included my work in her superb publication, which expanded my field of vision and got me to reread Ursula K. Le Guin's masterful forest novella; and Jennifer Sigler, who gave me an opportunity to publish a text pondering the curious links between plant and human communities in *Harvard Design Magazine*.

Some of the archival materials on which the central ideas of this work rest were discovered in the expansive archives of the Canadian Centre for Architecture in Montreal, to which I was invited to develop an exhibition. I should thank Mirko Zardini and Giovanna Borasi for seeing the potential of a rudimentary project, Fabrizio Gallanti and Emilie Retailleau for the rigorous process through which it turned into an exhibition, and Phyllis Lambert for her comments and inspiration. As the project developed into a large-scale exhibition at the Nieuwe Instituut in Rotterdam, it gave me the opportunity to expand the research into economic and colonial realms. I thank Guus Beumer for the invitation, and Wendel Ten Arve and Babette Zijlstra for its realization. I am also indebted to Toon Koehorst and Jannetje in 't Veld who designed the exhibition and complemented my knowledge with their incredible curiosity for forest-related resources.

The frameworks through which to think about the subject were much developed through a series of exchanges with students and faculty at various academic institutions. I wish to thank Brigitte Shim and Robert Wright at the University of Toronto, who invited me to give a workshop during which I benefited greatly from the acumen of their students; Stephanie Seymour, participating in that same seminar, who most probably inadvertently opened my eyes to issues and complexities of indigenous forest stewardship; Kate John-Alder, who asked me to present the project in Rutgers and whose writings on Aldo Leopold were inspiring and instructive for parts of my work; Andrea Trimarchi and Simone Farresin, who invited me to the inaugural Prada Frames conference, which included a stellar group of presenters that filled missing pieces in the forest puzzle; and Celina Martinez, Carla Ferrer, and Thomas Hildebrand, who invited me to contribute to their comprehensive volume on the uses and prospects of forests in architecture and later to focus on these and other issues with an incredible forward-looking group of students during a summer school they organized at Zurich University of Applied Sciences.

Archival and fieldwork would have provided only partial accounts of some of the projects and histories I trace in this work had I not been able to speak to some of the people involved in their creation. I must thank the brilliant minds I was fortunate to interview for this project: Jacques Herzog, Jerry Franklin, and Peter Walker provided honest and insightful reflections on their work. Alan Sonfist and Meg Lowman described life-long forest experiences that shaped their views of the world.

This book would not have been possible without the following individuals: Kate Schell, who saw its merit and accepted it for publication; Sarah Rae, who oversaw the editorial process; Selena Hostetler who saw that it would reach completion on schedule; Lekshmi Priya, who oversaw its production and Wendy Jo Dymond, who copyedited the manuscript; the reviewers whose feedback shaped and refined the structure the book; Keren Kuenberg, who opened many avenues while doing research for this book; Mauricio Quirós Pacheco, who was there in so many key points to help this rolling project take new forms; Anthony Acciavatti, peer and confidant whose perspective and insight were invaluable; and, finally, dispensing of the one forest metaphor I allowed myself in this list, I thank Gali, Lily, and Ben, members of my own sacred grove.

INTRODUCTION

My safe place is a forest. At least that is how I keep recounting the forest stand I stumbled on one summer day in Montreal, which keeps coming back to me since. I was going with my wife to see the Parc du Mont-Royal, the landscape jewel crowning the city's esteemed piece of rock topography. Heading from the nearby Outremont neighborhood, we were taking advantage of the sunny morning to make our way by foot up the hill. As we meandered toward the direction of the parc, loosely relying on a map, generous suburban lots anchored by early twentieth-century granite homes gave way to a plain asphalt road whose one side was entirely domesticated while the other rose up to form a steep slope and a wall of trees – steep but seemingly navigable – so we decided to cut through in order to avoid the heat that was beginning to accumulate. Into the green wall we went, and there we were: in a forest.

I am well aware that such a beginning sounds familiar. But this was no *selva obscura*, and we didn't need a Virgil to guide us through. It was simply the most exhilarating place I have ever found myself in, and its magnificence was amplified by way of its sudden introduction. After a brief moment of low bush, we came upon an area densely planted with even-aged, light-colored maples. The proximity of the trees to one another sent them up looking for light, forming a vertical composition of grey strokes on a fuzzy green background. Crepuscular rays from above justified their reputation as "god's rays" by lending the scene a somewhat unearthly atmosphere. Our walking figures wandered strangely out of scale, and all ability to assess distance or speed was diminished. But this was not merely a visual intoxication. The temperature dropped, and the change in heat and humidity made our bodies feel at ease. Acoustics changed too, and our voices were muffled by the soft ground surface and countless sound-diffusing elements that adorned the scene. It was a deeply visceral experience that felt completely out of time, like a shuttle floating in green space.

I took a couple of photos with my Leica (this was in my pre-smartphone era), knowing that the attempt to capture an adequate visual imprint of this was probably futile but trying nonetheless with the optimism of an eager

1

DOI: 10.4324/9781003473411-1

tourist. Later I found that, aptly enough, none of the photos survived due to a clandestine malfunction of the camera. So there are no reproductions of this experience, visual or other – when I asked, my wife seemed to forget all about this place. It exists solely in my mind as a pristine memory of wonder, discovery, tranquility, delight, and other sensations frequently associated with leaving the city behind and venturing into Nature.

Of course, it was not. My entire forest sortie, immersive as it felt, was simply a set constructed of natural elements, planned to the minute detail by famed landscape architecture Fredrick Law Olmsted as a means to intensify the contrast between what was to be on the top – a generous leisurely civic space and the surrounding slopes that were meant to be mantled in green. So how come I was fooled by this unassuming ensemble of trees to feel like I landed out of culture? Ignorance surely cannot be blamed here, as I was familiar enough at that point with the history of North American forests to know they were mostly man-made. In fact, my excursion to Mont-Royal, besides being an enjoyable weekend activity, was part of a voluntary mission to visit every Olmsted project I could, as I considered him the godfather of American forestry. Never mind also that a short climb brought us back to an asphalt road and that, minutes later, we were strolling the wide carriageways among the belvederes and the Gilded Age chalet that overlooks the city, a true fin de siècle setting that could not be more civilized an experience. The proximity of this middle-class fantasy grounds to "my" forest did not hold in my mind. Neither did any rationalization of the place I was briefly introduced to: clearly it was only a minor appendix to the park; the trees were too young to be part of the original plan, and the plan was not executed anyway in its original form.

All this didn't matter. What I knew then and what I know now did not prevent a modest tree space from summoning other dark and not-so-dark forests that are nested in our collective imagination and individual psyche, places in which you can get lost even in broad daylight. The experience made me see more clearly the special connection humans maintain with forest spaces – something that may exist, in some way, in their biological wiring many millennia after they have altered their wooded environments. At the same time, the small, planted patch of trees, transformed in a flash into *sylva incognita*, highlighted the neural networks that connect forest metaphors across cultures, geographies, and times.

★ ★ ★

As I ventured deeper into forests, I began to see them everywhere in my own habitat. Making appearances in landscape architecture, urban design, or architecture, forests were played out as projections of subconscious anxiety, as inspiration for organizational logic, or as literal insertions of groups of trees in and around buildings. Expanding the search to other disciplines, I came to realize that those involved with forests daily, either as scientists or as managers, shared many tropes with spatial designers. They were, in fact, speaking in similar terms: some associated the forest's ecological behavior with its form, others discussed forest systems as communities or cities, while still others developed theories of how forests can be redesigned to fit our age.

These kinships led to three hypotheses that guide this book: first, that most of the forests we know are designed; second, that forest sciences, forestry, and spatial design share metaphors in ways that make their histories worthwhile to contemplate in relation to one another; and, third, that while these metaphors may appear vague in some expressions, they are, in fact, structuring principles that can be clearly traced. By following these hypotheses, this project became an attempt to circumscribe a cultural history of designed forests.

★ ★ ★

It is hardly a provocation to argue that forest environments are the products of human design. If earlier accounts could speak of primordial settings – *urwalds* that exist somehow out of reach, the charting of most of the earth's surface that is completed in the latter decades of the twentieth century makes such forests mostly the stuff of fiction. Historical evidence shows that environments that play the role of emblematic untouched forests in the popular imagination, from the Sundarbans mangroves in the Bay of Bengal to the Black Forest in southwest Germany, have long been managed and altered by people. Environmental historian William Cronon proposed that Native Americans in precolonial New England dramatically transformed forest habitats by extensive and systematic burning.[1] More recent scientific evidence suggests that pre-Columbian planting is responsible for current-day Amazonian tree species composition.[2] The mere notion of the Amazon as an untouched environment was dispelled by scholars since the 1980s, with prominent Brazilian geographer Bertha Becker referring to it as an "urbanized forest".[3] Even under a strict definition of design as a deliberate set of actions with clear organizational and formal prospects in mind, many forests display the traces of intentional human action that shape almost everything, from boundary lines to the structure of cells, thus making sylvan habitats the product of people's plans, ambitions, and fantasies. And yet Nature, like in my humble encounter in Mont Royal, is never too far out of mind. This inherent ambivalence asks us to pause and consider what really makes a forest.

A forest is not a rock or a cloud – it is not a thing in the world. Nobody is born with the innate ability to call out the forest from the trees: Is there a minimum number of trees that is essential in order to qualify as one? Does size matter, or is the Sihlwald and the Monteverde Cloud Forest the same in this regard? Does the type of trees have any effect on our definitions? Or do these definitions maybe depend on internal networks? Science tells us that trees communicate through subterranean mycelial highways; that trees may send nutrients to one another, shoot pheromones and aerosols into the air to signal danger, and break the wind to protect their fellows. But how complex should an ecology be to be ushered in the family of forests? The answers to these questions are unfathomable since forest is first and foremost a cultural definition. In many respects, its meaning has more to do with class distinctions than with the taxonomy of trees. Indeed, the English word comes from the medieval Latin *foris*, meaning "outside", which leads back to the original function of forests as game preserves for nobles, banned by decree to all but the higher echelons, their weapons, and their hound dogs. This distinction by legal and consequently spatial design was true in medieval Europe as it was in Mughal South Asia and in Qing China.

Thus, forests, both as an idea and as real environments, were designed by human activities and human intentions since time immemorial. In one of the most moving forest scenes captured by an artist, *The Hunt in the Forest* by Paolo Uccello (ca. 1465), this idea manifests itself in full force. Uccello, who was, according to Giorgio Vasari, captivated by the attempt to achieve depth in his paintings, created a powerful one-point perspective that organizes the entire space of the painting and lends it a dramatic tone, amplified by the high-contrast night scene. While precise and full of detail, Uccello was not seeking to portray an actual scene: at the time hunts were unlikely to take place after dark, and none of the many characters in the painting appears to be carrying a light. Instead, the painting dives into the essence of a complex and symbolic social activity: the nobles on their horses with the crescents on the trappings, their footmen with their spears and decorated daggers, the hound chasing the game, the gazelles being surrounded in their last living moments, and, above all, the dark of the forest, with the glitter of gold flecks in the foliage. What is striking about this work from a forest perspective is how artificial the space is. Anyone familiar with wooded terrains could tell that trees do not grow in such a way when left to their own devices. And indeed, the painting shows, closer to the pictorial plane, some logs which indicate previous human activity. The hunt, then, follows a deliberate and tedious work of design. Of course, *design* was not the term used in the fifteenth century to describe these activities, and even *forestry* or *forest management* would only emerge later as defined fields of expertise, but the anonymous men who prepared the forest for the hunt to take place did not only cut trees. They were aware of the minutiae of this well-structured ceremonial social activity: of how riders move and at what height, of the spaces that are necessary for a chase to be possible, of the limits of the forest that should be invisible and yet defined. In their felling and cutting and clearing, they responded to these specifications and created a total space that was custom-made for a specific activity: it is an entirely designed forest (Figure 0.1).

Proposing that almost all forests are designed to an extent is not the same as suggesting that they have an agreed-on function within spatial design. Architecture specifically has led a fraught relationship with vegetal life. As architectural theorist Sylvia Lavin notes, the etymological homology between plants and architectural plans as *planta* and *pianta* is indicative of times in which "new buildings and living organisms, like young trees and seedlings, were implanted in the earth because that was where they were understood to grow and propagate... architecture was itself a plant".[4] However, around the time Uccello was completing his painting, things changed, and the discipline of architecture was enshrined as separate from the craft of the builder.[5] The abstractions of plans that were to become buildings, often drawn directly on the ground in preparation for construction, pushed actual plants off the site and out of mind. Almost six hundred years later, this fundamental resentment is safely in place. Notwithstanding the surge of environmental awareness and eco talk, the mainstream of architecture and urban design is still struggling to figure out whether trees are friends of enemies. A common saying among my graduate school peers was "When in doubt, plant a tree", inadvertently exposing the recklessness of professional disciplines that regard trees merely as objects to be added or subtracted, placing their trust solely in the power of numbers.

One possible explanation for this incapacity lies outside of design, within forestry itself. Rooted in the imperative to remove maximum wood material from the land and regrow its timber stock as swiftly as organically possible, forest management has been obsessed with metrics for centuries. While the vanguard of forest science and management has been laboring to expose the complexity of forest systems, the historic alliance with economic thinking is still so insidious, that even when trying to save the planet experts are sometimes aiming at the wrong target. Rosetta Elkin writes that "tree-planting fixations have become a political and industrial act rather than environmental necessity, exploiting the plant using persuasion, aggression, and control. This is not a projection; this is one of the world's most rehearsed spatial practices".[6] Foresters, she argues, continue to selectively determine value and focus on the more visible parts of forests because massive tree-planting initiatives are easier to fund and provide better photo ops. Rather than improving existing conditions, these projects

FIGURE 0.1 *Hunt in the Forest*, Paolo Uccello, c. 1465. Public Domain.
Painting in the collection of the Ashmolean Museum, University
of Oxford.

lead to the disappearance of complex ecologies and to "commercializing
plant life". This line of thinking flows into adjacent disciplines. Elkin
warns that the design professions "are complicit in repeating an exhausted
association with forestation".[7] By doing so, they are doomed to continue
along the path of "merciless" tearing-up that is embedded in the logic of
forestry. In other words, designers are too often either ignorant or content
with the idea of treating trees as interchangeable units, thereby facilitating
the relentless transformation of sites on which their disciplines depend.

Beyond disciplinary critique, these observations point to the unpro-
nounced links that bind science, management, and design through shared
ideas. These metaphors, referred to as analogies or organizing principles
depending on the discipline, guide the work of forest ecologists as much
as they drive the aesthetics of designed buildings and underlie the development of economic theories as much as they are applied in the routine
management of commercial tree plantations. Untangling these ideas is a
task that demands moving across time and space. In recent decades, forests
have been cast many roles in the unfolding play of human history, reaching new dramatic heights with the looming omens of the climate crisis.

Forests are concurrently portrayed as allies in the struggle against carbon emissions, as exemplars of sound economic management, and as allegories to the point of no return reached by the hubris of humankind. These roles are mediated by metaphors that circulate within and between fields of expertise and between cultures, circumscribing how people think of and act on forests. This book situates these metaphors within the specific historical, social, and economic contexts in which they were shaped and traces their sinuous evolution and elusive resonance in thoughts and ideas that lead up to the current moment.

Venturing into the cloud of forest metaphor reveals a plethora of ideas that exist in various levels of generalization. Take for instance the notion that forests and civilizations are diametrically opposed, an idea that some-how survives well into the twenty-first century. Robert Pogue Harrison, professor of Italian literature, argued that the forest prefigured human settlement both as historical fact and as myth of origin. "Western civiliza-tion", he writes, "literally cleared its space in the midst of forests. A sylvan fringe of darkness defined the limits of its cultivation, the margins of its cities, the boundaries of its institutional domain; but also the extravagance of its imagination."[8] This extravagance is often not so imaginative, and the regarding of forests as something of an alter ego of human cultures leads to simplified accounts of their relationships. While it is conventional wisdom that human actions are responsible for the troubling damage to forests worldwide, the actual dynamics of forest cover tell a more nuanced story. Recent years brought thousands of forest fires resulting from slash-and-burn operations. These fires have consumed vast areas of rainforest, sending smoke thousands of kilometers away to darken the skies of São Paulo and even prompting usually apathetic global leaders to take action. Indeed, since the 1970s, when the Trans-Amazonian Highway provided access to what was referred to as *infierno verde* (green hell), deforestation drastically reduced the area of the rainforest. However, a close competitor in tree killing statistics over the past two decades is not the bulldozer but *Dendroctonus ponderosae*, the mountain pine beetle that blitzkrieged its way through North American forests, leaving major destruction in its wake. This outbreak is assumed to be a result of global warming, as the beetle has a longer breeding season, which affects its proliferation. Simultaneously, scientists project that in less than fifty years from now, boreal forests will migrate north under the aegis of climate change and cover the tundra in Alaska and Northern Canada.[9] This afforestation is far from welcome, as it is inferior in its solar radiation reflecting capacity when compared to snowy surfaces, thus contributing to global warming.[10] In the opposite direction of the arrow of history, researchers concluded that the South Pole was once covered with temperate rainforests during the mid-Cretaceous period when global warming reached an all-time peak, long before the human race entered the picture.[11]

The assignment of absolute positive value to planting that Elkin refers to is another idea that is couched within the broader metaphor of a return of Nature. Cronon, in a critical account of the myth of wilderness, associated this ambition with "the illusion that we can somehow wipe clean the slate of our past and return to the *tabula rasa* that supposedly existed before we began to leave our marks on the world".[12] The religious and moral undertones of such conceptions can be found in countless accounts, scientific analyses, and design projects related to or taking inspiration from forests. But the reach of these ideas is so widespread and so diffused as to render them something of a constant presence in various cultures, times, and geographies.

As this work progressed, it became evident that it should steer away from these generalities and concentrate instead on a set of specific forest metaphors. The exploration of these metaphors ultimately shaped the structure of this book. While they are not entirely mutually exclusive, their boundaries are distinct enough to be considered independent of one another. Working in such a way, I often found that it challenged common distinctions within historical analysis. The deliberate design of forest environments, particularly on a large and systematic scale, is frequently associated with colonial conquests and their extraction economies, with sharp lines drawn between Western and non-Western civilizations, and with capitalist imperialism that undermines traditional ways of living. Delving into the histories of specific forest metaphors allowed for more complex narratives to be sketched out. It facilitated for instance a retelling of the work of a German forester in British India who not only promoted the extractive ideology of empire but battled against it, empowering local expertise that leads to a first-of-its-kind large-scale forest plan and subsequently transforming forests back in the metropoles; of a president in Vietnam that wages war on his country's forests in an attempt to associate them with antagonizing social groups he seeks to contain; of scientists and activists working in the highlands of Ethiopia, applying high-scientific ideas to connect distinct forest forms around orthodox churches with ecological and spiritual performance; of a Chinese communist leader taking cues from American experiments in large-scale planting, which were inspired by earlier Russian attempts to define national identity through environmental alteration; of legal scholars conversing with indigenous worldviews and spiritual writings when discussing the boundaries of rights for nonhuman species; and of cyberneticists and computer scientists, seeking intelligence in a communion of plants and machines.

Each of the chapters in this book follows a single forest metaphor, weaving a narrative that connects various routes and tributaries in its development. This narrative approach allows for the exploration of cultural contexts that drive contemporary ideas and projects related to designed forests. The chapters lead into one another, tracing a trajectory that oscillates between attempts to subdue forests and efforts to reconcile their overwhelming presence. This arc begins with a focus on the projection of reason onto the natural world, then moves into the exploration of the functions of forests as unknowable environments, and continues by looking at experiments of achieving stability under updated definitions of equilibrium and then follows ambitions to introduce ecological complexities into buildings and cities, and finally to hone in on influential concepts of intelligence that envision the bringing together of plants and humans in transpersonal forest environments.

The first chapter, "Engineered Forests", deals with centuries of attempts to rationalize forests as sites of production and as public displays of advanced organization. It begins by unpacking the abundance of forest metaphors used by spatial designers in varying geographical contexts and site conditions to describe overtly artificial edifices. Tracing a surge in the usage of such metaphors over the last few decades, the chapter historicizes it in relation to the growing shift away from parks to forests as leading corporate metaphors for controlled nature. The chapter then explores the origins of the idea of rationalized forests, stemming from the meeting points between extraction economies in the colonies and Enlightenment science. It moves on to the concurrent beginnings of landscape design and professional forestry in the United States during the late nineteenth century, focusing on the work of Frederick Law Olmsted in establishing the plan for the Biltmore Estate and its utilization in making the case for widespread scientific forest management. Finally, the chapter explores the logical conclusion of these ideas by focusing on the Weyerhaeuser forest company and its introduction of experimental designed forests as part of its public campaigns. It culminates by looking at the idiosyncratic Weyerhaeuser headquarters building, designed by SOM San Francisco and Sasaki, Walker, and Associates, bringing to perfection the idea of designed environments that represent and shape a radical reinvention of nature.

The second chapter, "Jungle", begins by sketching out the limits of the rationalizing drive discussed in the previous chapter. It follows accounts of unruly primordial forests challenging human physical and mental capacity. The retracing of such accounts in travel literature and scientific reports of the eighteenth and nineteenth centuries fleshes out the undertones of the jungle metaphor, representing not only physical duress but also moral and intellectual degeneration. The chapter then explores how these characterizations, often piggybacked on colonial expansion and the constant effort to draw a line between "civilized" and "uncivilized" peoples, informed the work of forest scientists dispatched in British India to rationalize its forest production yet found themselves dumbfounded by the complexity and foreignness of Indian forests. The chapter follows the progress of this work as it informed the creation of the first large-scale forest plans, crafted first by teams of Indian and European foresters, and then transforming the centers of colonial power. This work outlined the discipline of forestry as it shadowed other imperial campaigns in the British and then American cases, extending well into the twentieth century when the jungle looms large in the American psyche following its encounters in Southeast Asia. The chapter then delves into the dramatic large-scale forest redesign project during the Vietnam War, which blurs the lines between common distinctions made between East and West, developed and underdeveloped, and colonized and indigenous. With that, the chapter manifests the metaphor of the jungle as a means to exert social distinctions between "us" and "them."

The third chapter, "The Thousand-Year Forest", explores the idea of forests as harbingers and symbols of stability. Recounting the influential economic work of early nineteenth-century writer Johann Heinrich von Thünen, it establishes a link between notions of clear forest form and those of economic and hence social stability. Von Thünen's ideas about forests are understood not as static land-use allocations but rather as dynamic constructs that can react to changing conditions while maintaining internal stability. Similar ideas are at work in twentieth-century regional schemes that attempt to marry forestlands, industrial production, and community development over the long term. The chapter investigates far-reaching experiments in the context of the American New Deal, which brought together private and public interests to a point bordering socialist schemes. It then traces the reemergence of stability as a defining concept in ecological sciences during the 1950s and shows the complex interactions of forest forms, ecological equilibrium, and social stability, which take place in what becomes a vanguard region of scientific forestry during the 1970s: the Pacific Northwest. The chapter then traces the presence of the forest form as a performative component in two urban projects: OMA's Downsview Park in Toronto and Herzog & deMeuron's master plan for the expo grounds in Hannover. It concludes by tracing contemporary experiments to associate form, ecology, spiritual, and social stability in the highland forest churches of Ethiopia.

The fourth chapter, "Ecological Havens", expands on the wishful ambition of spatial designers to build on the combination of environmental crises and ecological concepts to introduce forests into their architectural, urban, or regional schemes. Focusing on the specific agenda of Ecological Urbanism, the chapter contextualizes the mushrooming of forest buildings around the world, from Emilio Ambasz's ACROS building in Fukuoka to Ken Yeang's tropical high-rise projects in Malaysia and Singapore and Stefano Boeri's Bosco Verticale, tracing their origins in 1970s calls for the greening of architecture. These calls, often responding more to internal disciplinary feuds than to real-world challenges, set the blueprint for contemporary arguments that display merely a superficial understanding of forest life. The chapter then follows the development of urban forestry, which, while influential worldwide, often displays similar shortcomings. It then traces how the metaphor of complex ecologies has been applied in projects such as Alan Sonfist's Time Landscape and the Wildlife Conservation Society's Mannhatta Project, which sought to recover and largely reinvent precolonial pasts and thus undo the evils of industrial civilizations, thereby connecting to an ecological sublime.

The final chapter, "Ubiquitous Intelligence", probes the ways in which the idea of nonhuman intelligence permeates thinking about natural systems, eventually making its way to reconceptualizing forests as sites of advanced interactions between humans, plants, and machines. Beginning with Terra0, a prototype for a self-managing forest in Germany, the chapter traces the sea change in legal thinking that allows for the endowment of rights to nonhumans, based on ideas from indigenous cultures. It then leads into the complex territories of nonhuman sentience, charting the resonance and expression of these ideas in two design forests: Eero Saarinen, Kevin Roche, and Charles and Ray Eames's IBM Pavilion at the New York World's Fair and Cedric Price's unrealized Generator project in Yulee, Florida. It then traces the next phase in the evolution of such hypotheses in the ubiquitous computing movement, and its applications in natural settings as multi-sensor networks that evoke notions of thinking forests. The chapter then concludes by pointing to the curious convergence of machine and plant sentience in forest environments, where the intelligence metaphor is ultimately applied to represent the metaphysical frontier of humanity as we know it.

The lessons of a cultural history of designed forests, or of any history for that matter, cannot be foretold. However, I do hope that by tracing the long-standing sharing of ideas between the science, management, and design of forests, a kinship between these disciplines is outlined that can be useful in facing fundamental human challenges brought about by environmental crises. More specifically, this history also bears meaning for the present situation in which spatial design finds itself. Having been for so long the accomplice of capital and development in the pollution of the environment and the depletion of planetary resources, it finds itself in an uneasy position: Is it even moral, knowing what it knows today, to continue practicing architecture, landscape architecture, or urban design as before? On the surface, the forest provides a tempting way out of these predicaments: if design could only become a forest, nobody would be able to question its motivations. This book tacitly suggests that in hiding behind the forest veil, spatial design not only rehearses the shortcomings of previous generations but also gives up its own patrimony, which includes the ability to reconsider resource extraction and rework the relationships between humans and environments. Once this patrimony is critically considered, it may equip designers with the tools to realign their mission in a world that is going up in flames.

Notes

1 Cronon, William. *Changes in the Land: Indians, Colonists, and the Ecology of New England.* 1st ed. New York: Hill and Wang, 1983, 30. This has since been substantiated by palaeoecological studies, even though the precise extent of the impact of indigenous ecological alterations is still being debated. See Abrams, Marc D. "Fire and the Development of Oak Forests." *BioScience* 42, no. 5 (1992): 346; Oswald, Wyatt, David Foster, Bryan Shuman, et al. "Conservation Implications of Limited Native American Impacts in Pre-Contact New England." *Nature Sustainability* 3 (2020): 241.

2 Levis, Carolina, Flavia R. C. Costa, Frans Bongers, Marielos Peña-Claros, et al. "Persistent Effects of Pre-Columbian Plant Domestication on Amazonian Forest Composition." *Science* 355, no. 6328 (March 3, 2017): 925–31.

3 Published in English in Becker, Bertha. "Undoing Myths: The Amazon – An Urbanized Forest." In *Brazilian Perspectives on Sustainable Development of the Amazon Region*, edited by Miguel Clüsener-Gogt and Ignacy Sachs, 53–89. Paris: UNESCO, 1995.

4 Lavin, Sylvia. "Reclaiming Plant Architecture." *e-flux Architecture*, August 2019. https://www.e-flux.com/architecture/positions/280202/reclaiming-plant-architecture/.

5 This gradual process received a notable thrust through the writings of Leon Battista Alberti, whose treatises wielded great influence over many generations of architects.

6 Elkin, Rosetta. "The Prefixes of Forestation." In *Harvard Design Magazine* 45 (Spring/Summer 2018): Into the Woods, 10.

7 Elkin, Rosetta. "The Prefixes of Forestation." In *Harvard Design Magazine* 45 (Spring/Summer 2018): Into the Woods, 6.

8 Harrison, Robert Pogue. *Forests: The Shadow of Civilization.* Chicago: University of Chicago Press, 1992, ix.

9 Rotbarth, Ronny, Egbert H. Van Nes, Marten Scheffer, et al. "Northern Expansion Is Not Compensating for Southern Declines in North American Boreal Forests." *Nature Communications* 14, no. 3373 (2023).

10 Hasler, Natalia, Christopher A. Williams, Vanessa Carrasco Denney, et al. "Accounting for Albedo Change to Identify Climate-Positive Tree Cover Restoration." *Nature Communications* 15, no. 2275 (2024).

11 Klages, Johann P., Ulrich Salzmann, Torsten Bickert, et al. "Temperate Rainforests Near the South Pole During Peak Cretaceous Warmth." *Nature* 580 (2020): 81–6.

12 Cronon, William. "The Trouble with Wilderness; Or, Getting Back to the Wrong Nature." In *Uncommon Ground: Rethinking the Human Place in Nature.* New York: W.W. Norton, 1995, 80.

1

ENGINEERED FORESTS

Forest metaphors abound in architecture. One doesn't need to dive too deep into the annals of Archdaily or Dezeen – two of the leading diffusers of architectural culture and imagery of our day – to come across one. In many cases, an array of vertical elements – concrete columns, wooden stilts, aluminum pipes – would be enough to qualify as a forest. Arguably, the mushrooming of metaphor relates in part to the function of text in these broadcasting channels, essentially a spacer between images, which cannot be expected to transcend the prose of a press release. But these bursts of Arcadia are transmitted equally from many parts of the world: they make an appearance in the most hyper-urbanized contexts and in the rarified countryside, in Asian boom cities, and in aging Western metropoles, in the financial hubs of the global North and new developments in Africa.

In Wuxi, a city of six million in the Jiangsu province, 365 angled steel columns supporting a golden canopy and surrounding a round theater building are described by the architect as a "bamboo forest", echoing the Sea of Bamboo Park that resides 100 kilometers away.[1] In Seoul, a community center in the Hannae neighborhood, an entirely urban affair made of glass and metal in the form of intersecting gabled volumes, is tasked with becoming an "artificial forest" and magically complement the trees growing in a nearby park.[2] In 2016, car brand MINI hired a designer to make polycarbonate boxes lined with houseplants and situated in the streets of London, labeling these unassuming translucent volumes "forest pavilions".[3] In 2019, at the Monsanto park next to Lisbon, architects placed 3,411 white-colored wooden posts in the ground, creating paths that lead to a small restaurant and a bar space. Hoping perhaps for a resurrection of the stumps back into living organisms, the project is called Floresta Branca (White Forest).[4] In Denver, a pneumatic nylon structure created as a temporary pavilion for the occasion of the 2008 Democratic

DOI: 10.4324/9781003473411-2

Party convention was described as "air forest".[5] While its bulky mass was more reminiscent of a plastic gorge, the architects insisted that it deserved its name by virtue of completing a gap in a ring of trees surrounding Ferril Lake. Neither the nineteenth-century gentleman who designed the park nor the acolytes of the convention gathering for their cocktails under the synthetic canopy could contradict that argument.

This profusion called for a more rigorous analysis of the forest metaphor as it threaded through design discourse. Examining the Avery Index for Architectural Periodicals, the term *forest* appears more than 2,300 times between 1870 and 2023.[6] The frequency appears to increase over the years, from sixty mentions in the 1970s, to 350 in the 1990s, and 680 in the 2010s. In order to better understand this general tendency, the data was combed differently. Focusing on publications specifically addressing architecture and urban design, rather than gardens or landscapes, the datasets were cleaned by removing place and designer names which included "Forest".[7] This refinement yielded 660 results, spanning from 1899 to 2023, which revealed a more nuanced trend. While the term almost doesn't appear in the first half of the twentieth century, the 1950s marked the beginning of a period of growth in mentions, which remains moderate until it visibly increases in the 1980s and then spikes in the 1990s and again in the 2010s. While the sample size and scientific protocols are still limited from a statistical standpoint, they do expose a substantial inflation of forest presence in spatial design discourse.

Even more telling than the numbers were the qualitative observations to be gathered from reading the content. In the 1950s and 1960s, the term *forest* was primarily used to describe the setting of architectural projects (i.e. "Alvar Aalto in the Finnish forests").[8] during the 1970s, it becomes more plausible to trace metaphorical usages in describing the ideas behind projects (for instance "Forest Murmurs" to describe a college music building).[9] The 1980s witnessed more poetic usages of the term ("Pine Forest Palette: Rustic Storybook Cabin High in the Rocky Mountains" or "A Machine in the Forest: Center for Innovative Technology, Herndon, VA").[10] During the boom of the 1990s, *sustainability* and *green design* are often used as well in relation to forest, with the *forest* term frequently occupying article titles ("A Forest Runs through It", describing an airport building in Kuala Lumpur).[11] In the second boom of the 2010s, ecological and environmental anxieties are integrated more frequently in relation to the forest (titles include "Paradise Lost" and "The Jungle's Call").[12] Even more important, entire issues are dedicated to forests, exposing an understanding that forests encompass both technical and cultural aspects pertinent to the design professions.

***

In late 2023, Oliver Wainwright, architecture critic for The Guardian, posted on social media images from his visit to Thomas Heatherwick's 1000 Trees Mall in Shanghai. The caption "I climbed to the top of [the project] so you don't have to" betrayed the critic's dismay at what he found in that building, including "colonic elevator buttons" and a "Heatherwick shrine on the ground floor". Such idiosyncrasies aside, the project can be better understood as echoing a deeper desire to naturalize a building by way of metaphor, and, more specifically, by summoning trees to the rescue. The 1000 Trees project, contrary to what its name suggests, is first and foremost concrete: a 315,987-square-meter articulated mass, shaped into two topographical volumes, only one of which is currently built. The project is a mixed-use development, bringing together high-end fashion brands, cafes and restaurants, commissioned public art, and housing units. Its vast interiors are punctuated by a large number of concrete columns arranged in a simple 9-meter grid, which received a great deal of attention from the architects. In the design process, they developed a ripple pattern to adorn the concrete surface of these columns with what appears like concrete rope wrapped around each. The hundreds of them were produced in various heights to support the mountainous silhouette the architects had in mind, and planters were put on top. Notably, this rather straightforward design development is presented by the architects in highly metaphoric terms. The project, they pose, is "designed to emerge like a pair of forest-capped mountains from the waterfront in Shanghai". The grid of columns is "an exoskeleton … revealed to liberate the internal space and highlight the structural means to lift up a landscape, tree by tree".[13] The planters they see not just as vessels but also as an infrastructure for a new forest to grow: "It is as if green shoots have sprouted up through the building to bloom on the skyline – the top of each column extends into a broad planter, bringing nature close to each level and every terrace." The descriptions highlight the innovative aspects of the design, but upon closer examination, the project resonate architectural attempts from previous decades, which sought to grapple with similar programmatic content and reconcile trees and capital within buildings.

***

The challenge of integrating actual plants inside a building is as old at least as the Roman *specularia*, described by Pliny the Elder in the first century CE. In his *Historia Naturalis*, Pliny describes how the emperor Tiberius's unsatiable appetite for long melons was supplied in his Villa

Jovis in Capri by "beds mounted on wheels which they moved out into the sun and then on wintry days withdrew under the cover of frames glazed with transparent stone".[14] And so the proto greenhouse, diverging from agricultural methods of crop protection, was contingent upon material abundance and the workings of power from its early appearances.[15] This alliance is threaded throughout the history of orangeries in Europe, mostly constructed on the grounds of lavish estates for the consumption of citrus. In the early sixteenth century, Italian species were supplemented by orange trees brought from India by the Portuguese, who were then at work transforming the ecologies of Madeira and the Canary Islands with citrus and sugar plantations. The exotic gradually became commonplace with the famous five – orange and lemon trees, myrtles, pomegranates, and oleanders – adorning the lands of the wealthy. During the eighteenth and nineteenth centuries, trade with the colonies coalesced with building technologies to form modern greenhouses, in which glass and iron substituted traditional masonry and wood, becoming usable year-round, architecturally distinct structures.[16] Victorian-era literature, from Charles Dickens to Oscar Wilde, is ripe with descriptions of domestic greenhouses as symbols of privilege, neglect, and eccentricity. At the same time, similar architectures were utilized outside of the private sphere, in a plethora of botanical gardens constructed in the colonies as in the metropoles. These hubs concentrated scientific research in centralized facilities under the aegis of state power, usually for the purpose of their economic development. The first decades of the twentieth century brought the decline of the domestic greenhouse, which became too costly to maintain and, as a backlash to its previous cultural status, went out of fashion as an ornate decorative relic of another era. In the United States, Frank Lloyd Wright, always with a foot in the nineteenth century, designed several conservatories for wealthy clients throughout his career, from the axial one at the Darwin Martin House in Buffalo (1906) to his circular conservatory at the Tirranna house in New Canaan (extension completed in 1958). A decade later, Kevin Roche proposed a small orangerie for a private client in Columbus, Indiana (1968, unbuilt), and Paul Rudolph designed a glass conservatory as part of his angular Green residence in Honesdale, Pennsylvania (1972). But these were exceptions, and most of the developments in greenhouse building technology and environmental control were led by research institutes in universities or other public agencies (Figure 1.1).[17]

In architecture, these developments were echoed by often futuristic designs of structures that were able to regulate heat, airflow, and humidity, based on the conviction that technology could answer the rising concerns regarding pollution and other forms of environmental degradation.

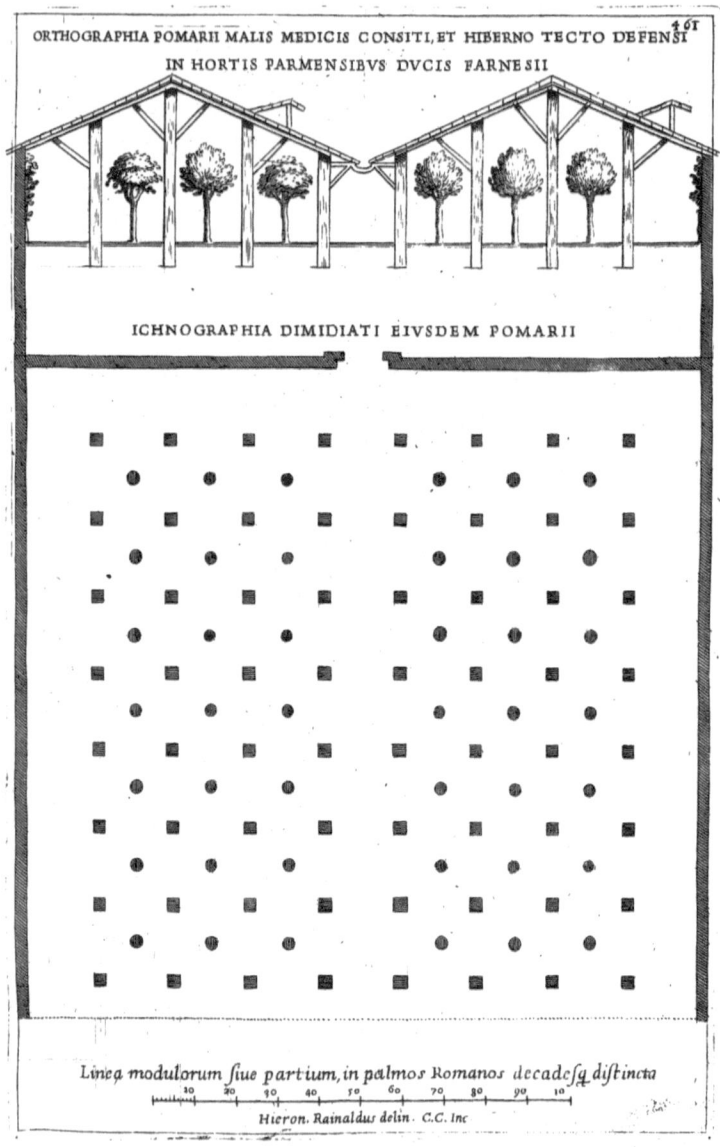

FIGURE 1.1   Design of a permanent structure for Duke Farnese's orange trees, from *Hesperides* by Giovanni Baptista Ferrari, 1646.

Reyner Banham's essay "A Home Is Not a House" published in 1965, already critically assessed the particularly North American desire to cram technological paraphernalia in domestic environments to the extent that architecture itself becomes redundant.[18] Banham continued his inquiry into architectures based on man-made climatic control, published in his well-known book *The Architecture of the Well-tempered Environment*.[19] Contemporary proposed or built projects that followed the same thread presented fantastic, energy-intensive, hyper-architectural tempered environments, that could exist either as high-end research facilities or as eccentric demonstration projects.[20]

At the same time, there emerged the idea of the corporate terrarium – plant-full interiors designed and environmentally managed for the well-being of organization men. The forerunner of this genre was Kevin Roche and John Dinkeloo's Ford Foundation headquarters in New York, inaugurated in late 1967. The building, a twelve-story volume framed in exposed Cor-Ten steel and granite on 42nd Street in Midtown Manhattan, was laid out in a C-shaped plan, in which office spaces were organized around a ten-story high atrium with green terraces. The atrium was planted with southern magnolias, jacarandas, evergreen pear, and Japanese cryptomeria above various shrubs and bushes and ground cover plants. The fourth wall was made of glass, as was the skylight, inserting light into this unexpected green space. The scheme for the Ford Foundation presented a clear break from earlier Roche and Dinkeloo corporation headquarters buildings, created with or under the influence of Eero Saarinen. Projects such as the Bell Labs in Holmdel, New Jersey (1962), and John Deere Headquarters in Moline, Illinois (1964), demonstrated a clear dichotomy between an architectural object and the surrounding landscape, designed in both cases by Sasaki, Walker & Associates. In these projects, the greenery was conceived in an almost beaux-arts fashion as a stage, ceremonially highlighting the building for distinguished visitors, while employees often accessed their workplace via utilitarian passages. In contrast, the project in Midtown sought to bring the four hundred people who worked at the foundation into a constant proximity with natural elements (Figure 1.2).

An article in *Domus* magazine, published shortly after the building opened, hailed its ingenuity in providing a space that was "lush with vegetation and air conditioned", that is, natural and technological like a greenhouse. The writer also recognized the applicability of this architectural solution "completely new on this scale" in large cities, where "there is smog, rain, wind, and noise".[21] In other words, the building posed the possibility of isolation from the surrounding social, political, and environmental degradation then facing New York and other American cities.

FIGURE 1.2   Atrium of the Ford Foundation building, 2013. © David Leventi/
Courtesy Rick Wester Fine Art.

David Gissen, writing an environmental history of the project, highlights
the work of landscape architect Dan Kiley who, in the design of the
plant life in the building, borrowed concepts from Fritz Went, a Dutch
plant scientist who rose to prominence following his discovery of auxin
in plants in the 1920s.[22] Went was operating an experimental Phytotron
at CalTech since the late 1940s, testing the interactions between air pollution and plant growth. Therefore, the Ford Foundation building was
backed by advanced scientific and mechanical concepts orchestrated to
re-create an interior piece of nature. At the time, Roche regarded the
enclosure as a park with trees and greenery to enjoy throughout the year.
Later, he would refer to it not as an architectural whim but as the logical conclusion from numerous conversations with employees in the big
corporations he worked for, in which they would often pronounce a
desire to be closer to nature. "We have done so ...", he claimed, "in
several different locations by introducing greenery, based on the old Jung
theory that humans must connect to greenery or to nature."[23] The Ford
Foundation building and its wooded interior marked a moment in the
reinvention of nature for the corporate class. While still following the

hyper-engineered thrust of the period, it cultivated the link between financial power and forest metaphors. Ada Louis Huxtable called it a "shimmering Crystal Palace", which indeed it was, a twentieth-century greenhouse-type building, constructing an interior equivalent to that famous palace of leisure and economy that showcased the prodigious achievements of Empire,[24] only this time the world inside the building was made in the service of corporate desire.[25]

At the same time the Ford Foundation building opened, John Portman and Associates were at work on several major hospitality projects in declining American city centers. In these interiors, the hypothesis of a separate world for the citizens of corporate America achieved further refinement. They could traverse the bridges above Main Street, spend a weekend at a conference, breathe artificially circulated air, and sip a drink among the green planters decorating the booth before getting back to their cars and leaving a blighted Atlanta or San Francisco behind. From an architectural standpoint, the large mass, deep floor plans, terraced spaces, and intensive use of vegetation situate Heatherwick's project quite close to Portman's terrariums. Portman's mastering of massive buildings as containers for almost theatrical occurrences that result from the layering of circulation and the multi-programmatic arrangements of impressive volumes is echoed in the 1000 Trees project, including the articulation of the vertical elements and the emphasis on the design of elevators as conspicuous "people movers". 1000 Trees is also connected to Portman's legacy in its deliberate use of vegetation as a mediator between voluminous commercial interiors and organic natural settings. Portman notoriously positioned thousands of live plants in projects such as Hyatt Regency in Atlanta (1967) and Hyatt O'Hare in Chicago (1969), which soon became a major maintenance challenge, forcing the Hyatt cleaning staff to replace dead plants and laboriously dust the leaves of the living ones to allow photosynthesis to occur. The eventual failure of plants to survive was in fact a recurring feature of such ambitious "maintenance environments".[26] As one result, the plants in several of Portman's projects were substituted by artificial plants during the 1970s and 1980s. The Ford Foundation forest also never flourished as it was hoped for by Kiley, slowly deviating from its designer's vision as subtropical plants were planted instead of the original species. In 2019, it was entirely rehauled, bringing in Shady Lady black olives, jacarandas, and Amstel King ficus from Florida to create a new canopy and a new atrium ecology.[27] The results are yet to be evaluated.

The life and death of great terraria advanced a new metaphor for the corporate class: instead of a park, a forest. But it had to be a very specific kind of forest, one that could be maintained and controlled. The engineered forest came in as an apt metaphor to the ways in which capital sought to organize the natural environment for centuries, but even more urgently at the age of uncertainty brought about by the social and political upheavals of the late 1960s. However, the origins of engineered forests can be found much earlier, in the eighteenth century, as theories of forest management and engineering are being concurrently developed in European contexts. The development of these ideas rested upon the expanding "timber frontier" of early capitalism, in which rising mercantile empires encountered sylvan territories and transformed them into cash crops: first timber, then cereals, then sugar. Environmental historian Jason W. Moore writes that

> [i]n the rise of capitalism, sooner or later everything returned to the forest. Every decisive commodity sector in early capitalism – metallurgy, sugar cultivating, shipbuilding, construction- found lifeblood in the forest. Even cereal cultivation – one thinks of Polish grain flowing to seventeenth century Amsterdam – was bound up with forest clearance on a grand scale.[28]

More specifically, the works of writers such as John Evelyn, which became highly influential in Enlightenment circles, made direct connections between the imperial need for naval power and the redesign of the English territory through intensive tree planting.[29] These concerns flowed to France through the circles of the Jardin de Roi, who took upon themselves to translate available English literature into French and often took extensive study trips to England. The Jardin du Roi was established in the 1630s following the model of the Hortus Botanicus in Leiden, which by that time proved instrumental in fostering the links between scientific knowledge and imperial expansion. Yet it was different in its close connections with the state, which allowed scientists to benefit from the global reach of French colonial acquisitions and at times, as Richard Grove writes, "exercise their knowledge of the natural world in political terms".[30] In his seminal work on the development of environmental theories, Grove describes the circle of scientists affiliated with the Jardin, who begin to see close connections between forests, economy, and environment, drawn from and later projected back by their followers upon tropical contexts. Two of these, each making significant contributions to the study and science of forest management, were Henri Louis Duhamel du Monceau and Georges-Louis Leclerc, Comte de Buffon.[31] Du Monceau developed an interest from a more agricultural perspective than Buffon, even though the two collaborated on studies on the structure and growth of trees in the 1730s. His background and empirical observations were initially developed in his estate, which he turned into a model farm. These biographical details informed the practical approach taken in his writings and the close connections he saw between engineering and forest management.

Du Monceau never visited the colonies, but he must have been aware of how sugar plantations worked, as he contributed an entry on sugar for a volume of the Diderot and d'Alembert's *Encyclopédie*, published in 1765.[32] Beyond the physiognomic features of sugar cane, the entry describes the refining process at length and refers to the term used in "the territories" to describe parts of the machinery.[33] The article also links to an adjacent entry on *sucrerie* (sugar refinery), which describes in detail the organization of a plantation in a tropical setting, including the dimensions of cane lots and the number of enslaved people, and accompanied by a detailed engraving of such operation in Martinique.[34] Whether inspired in part by the plantation model or not, du Monceau's treatises on trees, timber, and forest management, devised as non-rhetorical practical manuals for the farmer or landowner, presented a powerful holistic, multi-scalar and economic perspective of the forest.[35] Their road map of how to improve the land, manipulate the structure of trees, and connect their different parts with specific market requirements was a radical reinvention of the forest landscape as a proto-industrial endeavor that is integrated in the economy (Figure 1.3).

The engineering of the forest that du Monceau outlines occurs before the planting of the tree in land preparation; during its growth through coppicing, pruning, and bending; and, after its felling, through manipulations that compensate for natural defects and further process the wood material. On the macro scale, the entire forest tract is redesigned from scratch for that purpose. On *des semis*, du Monceau includes a plan for a forest of 5,000–6,000 *arpent*, laid out as so to optimize drainage, fire control, and the collection of wood. The plan, which strangely echoes contemporary landscape design schemes, includes a central square for distribution that is adorned with an obelisk, as well as secondary distribution points, interconnected by an interlace of alleys and paths. The edge of the forest is planted with large elms in a checkerboard pattern, "like in the avenues".[36] And so a systematized, optimized, forest plantation receives a civilized frontage, purporting to integrate into the urban and economic life of society (Figure 1.4).

★★★

A civilized frontage was provided at another lavish country seat, the Biltmore Estate in Ashville, North Carolina. In the 1880s, George Washington Vanderbilt, grandson of "Commodore" Cornelius Vanderbilt

FIGURE 1.3   Illustration from *L'encyclopédie* accompanying the entry on sugar plantations, 1765. Courtesy of BnF.

FIGURE 1.4   Ideal forest plan, from Duhamel du Monceau's *Des semis et plantations des arbres*, 1760. Courtesy of BnF.

and one of the country's most affluent heirs, began purchasing farms and forests around Ashville to build his summer house, which he called "my little mountain escape". The little escape evolved into the largest private home in the United States, built by a thousand workers, to which Vanderbilt feverishly shipped decorative elements from all over Europe, and which eventually led to his financial ruin. The French-style chateau was situated within a 125,000-acre precinct, and Vanderbilt commissioned Frederick Law Olmsted to design the land.

Olmsted commenced work on the project in 1889, selecting the site for the main house, designed by Richard Morris Hunt, and designing the formal gardens and the 3-mile road leading to the house.[37] Faced with the challenging task of giving form to an area a hundred times bigger than his Emerald Necklace, Olmsted surveyed the land, many parts of which he found to be in deplorable condition and marked by the destructive patterns of generations of settlers. His conclusion was that the majority of the estate should be afforested, as ground conditions would support a large variety of trees and shrubs that would grow rapidly.[38] In a letter to Vanderbilt, written in July 1889, Olmsted described the project less as a Versailles than a working forest estate. "At this point", he writes,

> I am inclined to advise you to have in view the establishment and maintenance of an unbroken forest from the north to the south end of the estate, to extend from the east border, as a general rule, to the edge of the river bottom on the west, but with a "Park" to be taken out near the residence.[39]

This vision required extensive engineering of the forest, which Olmsted sets the roadmap to, and predicts that will become "finer than any natural forest ... finer than any other planted forest".[40] At the same time, it allowed for economic planning. In Olmsted's mind, proper forest management was also a way to offset the substantial costs of maintenance incurred over the long term.[41] Vanderbilt accepted most of these points, and Olmsted pursued his plan of establishing a forest by building a nursery on site that could supply the massive quantities of seedlings and young trees, and later by hiring the young Gifford Pinchot to oversee the management of 4,000 acres that could prove the economic benefits of a properly engineered forest. Beyond the technical challenges involved, Pinchot's experiment was consciously designed as a demonstrative landscape with substantial public relations benefits, as he and Olmsted ventured to present it at the 1893 Columbian Exposition in Chicago.[42] While the Biltmore Estate never fully realized the economic potential of

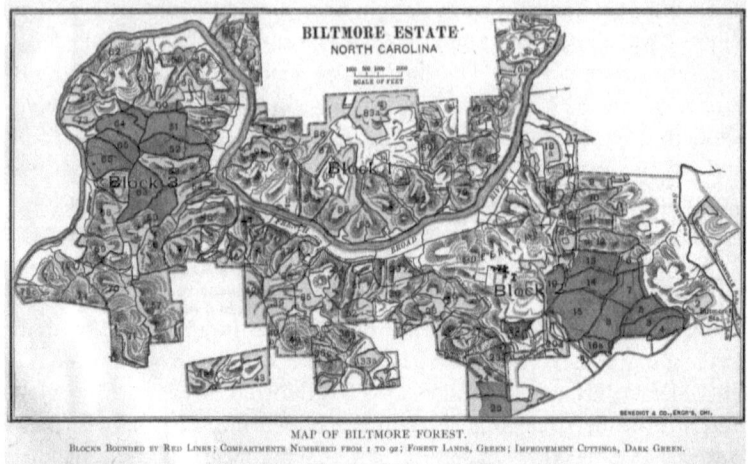

FIGURE 1.5   Map of Biltmore Forest, from Gifford Pinchot's report for the Columbian Exposition, 1893.

its advanced forest management, Pinchot reaped the benefits at the highly visible Columbian Exposition, which positioned him as a groundbreaking national expert. This reputation would in part lead him over the next two decades into the heart of the administration and into the American forestry hall of fame (Figure 1.5).

★★★

In the mid-1960s, forest products giant Weyerhaeuser Co. decided to follow other mammoth corporations and build its central headquarters on a forest site outside of Tacoma, Washington. The project, designed by Skidmore Owings and Merrill San Francisco with Sasaki, Walker, and Associates as landscape architects and inaugurated in 1971, propelled the metaphor of the engineered forest to new heights, including the well-designed frontage for forest regimes and intense economic activities: "A building that makes its own landscape"; "a dam"; "a horizontal skyscraper". These were some of the descriptions used in the architectural press in the attempt to pin down an apt comparison that could capture the essence of the project, which was quite surprising given the fact that its creators belonged to the high ranks of architectural production and that their previous work had been published extensively and in unequivocal terms by the same media outlets. This difficulty stemmed from the epochal nature of the project, which appeared seemingly out of nowhere and remained an exception in the portfolios of all those who were involved. Like many pivotal creations, it captured and reverberated the signals of a moment in time in unique and unrepeatable ways. It is indeed a strange building. A dark, extremely elongated mass, broken into five stepped floors covered in green, seemingly out of time to the point of flirting with ruin aesthetics, and certainly a pariah in the glowing and optimistic family of powerful headquarters buildings produced throughout the same decade (Figure 1.6).

While its siblings would typically be distinguished for their efficiency and innovation, the Weyerhaeuser building makes no attempt to position itself at the technological forefront of the building industry. This is most evident in its long facades, in which the glass walls are receded from the surface of concrete floor slabs. This achieves multiple effects: during the day the shaded glass becomes darker than the concrete, and the horizontality of the building seems to extend even further, turning it into a corporate version of Prairie-style fantasies. Combined with the surface treatment, mostly vertically lined rough concrete, it does make a visual reference to infrastructural developments one might encounter while driving on national highways. Its positioning, right on the valley floor and on top of a naturalistic pond created by a shallow creek that runs through the site, further enhanced its bridge-like impression, informing the "dam" descriptions. From an architectural history standpoint, it is tempting to see its physical presence in light of megastructural developments that were adorning the pages of architectural magazines. The timing seems right:

FIGURE 1.6   View of the Weyerhaeuser Headquarters, 1971. © Ezra Stoller | ESTO.

Reyner Banham would later declare 1964 as the year of the megastructure, Paolo Soleri was at work on his Valediga, which consisted of a large dam filled with the contents of an entire city, and Paul Rudolph was looking back to Ponte Vecchio in Florence when designing schemes for the Lower Manhattan Expressway Project and the Burroughs Wellcome building in North Carolina.[43] But these associations are unsubstantiated, and it is unlikely that they served as a major inspiration for the western forest company and its designers. Moreover, throwing the Weyerhaeuser building in the bag with these other projects takes away from its truly renegade treatment of the glass walls. The pushing back of these surfaces not only achieved visual impact but also allowed space for a surrounding walkway on each floor. These walkways were bordered by planters, and so one finds themselves in an ambiguous threshold, walking in a covered corridor but exposed to the elements; on the one side, seeing the forest through the ivy and, on the other, its reflection. Through the glass, the forest outside is amplified by lush, vegetated interiors. This ambiguity is no accident but the defining feature of the entire project.

It seems like every design decision manifested in this space works against the grain of corporate architecture as it was defined precisely by firms like SOM. Over the previous two decades, the model for such architecture was the flexible box enveloped by a glass curtain, and both the urban and exurban headquarters of burgeoning corporations – from the Lever Brothers to Pepsi-Cola, and from Bell Labs to IBM – followed and innovated within this paradigm. The long tail of the alpha curtain-wall projects was witnessed in many North American city centers as ubiquitous glass facades took over entire avenues and downtowns. Peter Blake, writing in 1965, vehemently attacked a newly built stretch along Sixth Avenue in New York, describing it as a "giant sample case of curtain wall salesman" and as part of the "slaughter that is our cities today".[44] Reinhold Martin, theorizing in retrospect the rise of the curtain wall as mass phenomena, associated its modular surfaces with the interior organization of spaces that were designed to "integrate the unpredictable needs of a rapidly changing workplace into an organized flux". In noting how Gordon Bunshaft and Nathalie de Bois of SOM organized the interior elements of the Union Carbide building in accordance with the architectural skin, Martin refers back to Bunshaft's observation of a period in which flexibility and non-specialized space played key roles.[45] It is therefore telling that Bunshaft was removed from overseeing the design of the Weyerhaeuser building after presenting to the client an object-like architectural scheme. Equally telling was the fact that the project was handed to Edward "Chuck" Bassett of the San Francisco branch as the new president of Weyerhaeuser deemed that task "more compatible" with a western architect.[46] The idiosyncrasies of the Tacoma headquarters are partly the result of an American-western-corporation-meets-American-western-designers process, topped with a refined and nuanced understanding of forests and forestry that infiltrate and shape the project in numerous ways. In the mid-1960s, Weyerhaeuser was at the helm of an industry that was undergoing dramatic changes and its president, George H. Weyerhaeuser, was well aware of the roles the headquarters building had in segueing his grandfather's conglomerate into the new era.

The Weyerhaeuser company was founded in 1900 by virtue of a massive land acquisition of 900,000 acres of forestland in the state of Washington that its founder, Frederick Weyerhaeuser, managed to acquire from railroad magnate James J. Hill, then president of the Great Northern Railway and majority owner of the Northern Pacific Railway.[47] Initially, the company sold standing timber, and focused on acquiring more land in the region. It opened its first mill in 1903 and gradually went into production and then boomed during World War I, which secured its status as an industry leader.[48] On the forest side, this meant intensive logging, with operations that consisted mostly of clear-cutting, which ensued large-scale ecological transformations in the region. Weyerhaeuser gradually began to implement advanced sustain-yield practices, establishing a research department and creating its own tree farms.

The Clemons Tree Farm, established by Weyerhaeuser in 1941 on a 120,000-acre tract of depleted forestland in Grays Harbour County, Washington, was a groundbreaking experiment in the artificial plantation of Douglas fir as it was in devising a strategy for shaping public opinion.[49] Like other timber companies, Weyerhaeuser had faced growing criticism of its logging operations, intensified by raging fires that consumed vast areas of forestland in the region in the 1930s, which changed government policies toward private forest management. The tree farm was understood as a vehicle to convince the public that the corporation had their best interest at heart.[50] As Emily Brock notes, the lumber industry sought to highlight parallels between privately owned tree farms and national forests as two sides of the same coin.[51] Weyerhaeuser invested much effort in portraying its industrial tree farms as rich ecological habitats where deer, eagles, and bears roam freely.[52] People were allowed to roam as well, as the company found it beneficial to use the tree farms not only as landscapes seen from the road but also as a living demonstration of the forest plantation's positive impacts on environment and society. Clemons was integrated into both the motorist's and the recreational tourist's itinerary through the creation of road maps and its connection with existing trails. Other tree farms proliferated, and by 1958, there were 140 of them throughout the country. The Pacific Northwest tree farm model, while ecologically debatable, outlined Weyerhaeuser's approach of using its sites as demonstrative pieces of a public relations strategy, hinged on direct interaction with visitors (Figure 1.7).

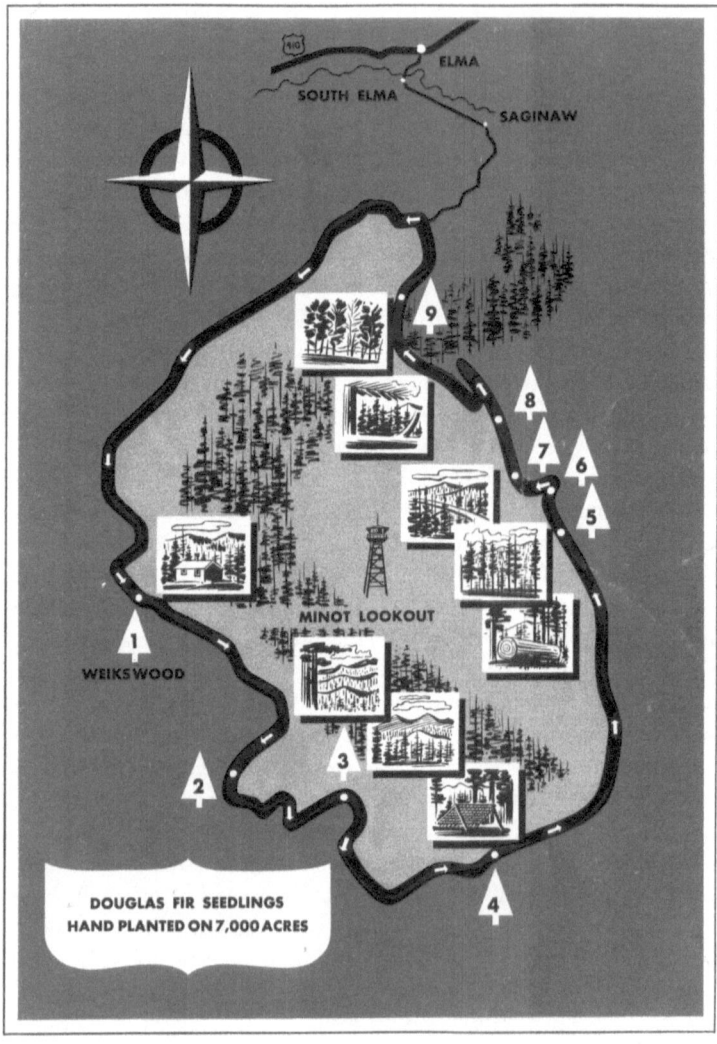

FIGURE 1.7   A visitor road map of the Clemons Tree Farm, 1940s. OHS digital
no. bb009968, Courtesy of Weyerhaeuser Company Archives.

The site for the headquarters building was initially selected by SOM out of four or five available options when Bunshaft was still leading the project. When Bassett took over, another site, the one on which the headquarters would end up being built, was agreed on. Walker recalls touring these forest sites with SOM and George Weyerhaeuser, having then a very basic understanding of forests and their management.[53] The site in Federal Way was chosen due to its connection to major highways and its high visibility from the surrounding roads. It was populated at the time by second-growth Douglas fir and red alder but was heavily disturbed. Bassett recalled that the land "had been cut up badly over a long period of time by county roads, abandoned access trails and spotty, semi-rural development", and yet, it had character that "immediately suggested a solution".[54] Walker's site analysis, which was encapsulated in a topographical model of the site, marked the beginning of an intense conceptual design phase involving both architects and landscape architects. At that moment, the teams prepared several schemes, placing the building on one of the two hills present on site. Then Bassett came up with the idea that he would later call a "simple statement.... Simply a number of terraced floors moving across from one side of a swale to the other".[55] From this point on, bold design decisions followed each other. Once the visibility of the building was established, the next big move was to get the parking out of the front, terracing the lots and positioning them perpendicular to the building mass. Then came in the idea of the "dam", creating two very different views towards the otherwise identical long elevations: from the south, it was seen through a meadow, while the north view positioned it behind a lake that reflects its image. "The forest", says Walker, "was supposed to fill in on the other sides."[56] However, the forest not only played the role of filler, but became an essential presence, informing design decisions on all scales: "the great panels of ivy [covering the stepped floors] were in the scale of the forest".[57]

This was in line with Bassett's idea of a building that would reach a point "where the landscaping and the building simply could not be separated, that they were each a creation of the other and so dependent that they could hardly have survived alone".[58] However, Walker's understanding of the project gradually moved beyond the immediate timescale leading to the project's completion to the pace imposed by the surrounding forest, leading him to refer to his work as "forest management ... more than landscape design".[59] The divergence is manifest in two key documents prepared during the design of the project: SOM's cross section, cutting through the middle of the building and showing the parking areas in elevation, and Sasaki, Walker, and Associates published landscape plan for the project, covering the entire site (Figures 1.8 and 1.9). The section

31

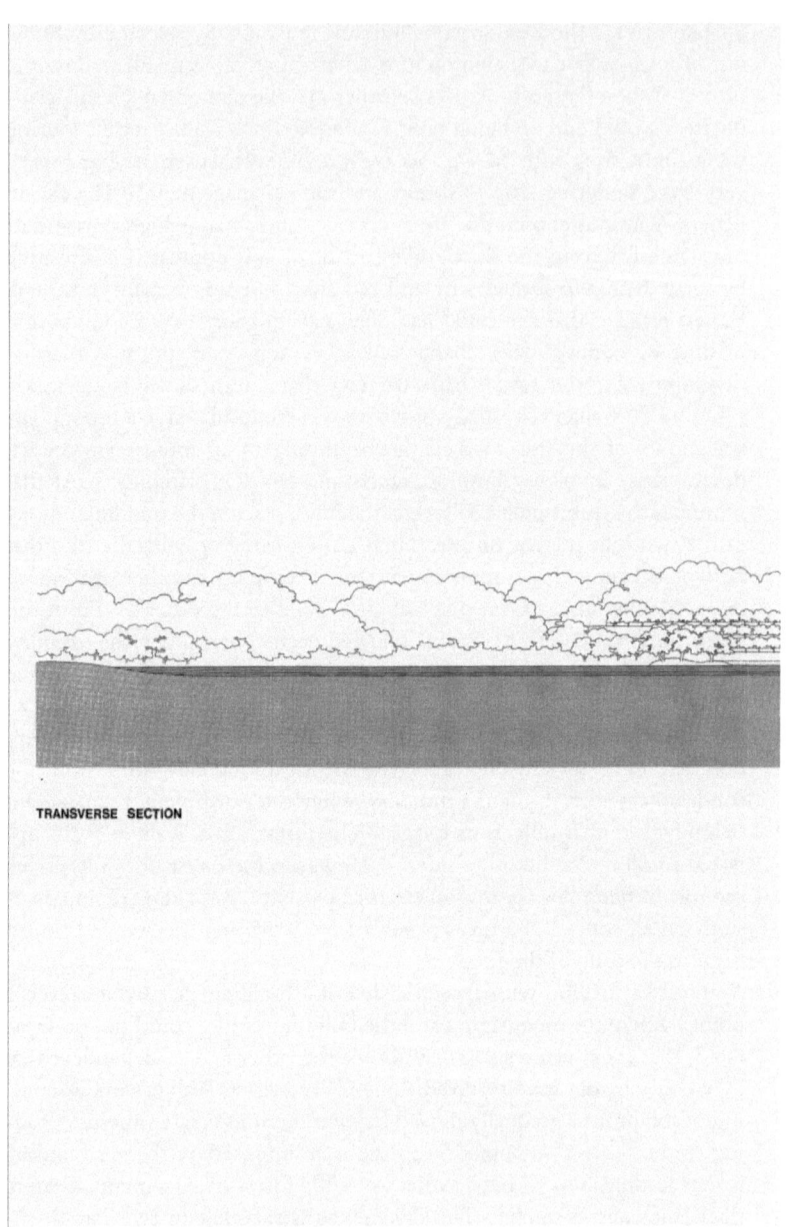

TRANSVERSE SECTION

FIGURE 1.8   Section of the Weyerhaeuser Headquarters, c. 1970. © SOM.

SOM
Skidmore, Owings & Merrill LLP

Weyerhaeuser Headquarters
Tacoma, Washington

FIGURE 1.9   Weyerhaeuser landscape plan, Sasaki, Walker and Associates, c. 1970. © PWP Landscape Architecture.

is a curious document, demonstrating the attempted immersion of the large building mass in the forest environment. Oriental in atmosphere, the building appears like a pagoda amid a billowy forest. The sycamore trees, planted in rows throughout the parking lots, seem like a plantation, and the entire scene is surrounded by an abstract canopy that has little to do with the actual trees on-site.[60] The landscape plan is far less poetic, demonstrating the idea of a dynamic and longer-term development of the forest site. It is especially striking when compared to earlier Sasaki, Walker and Associates schemes for campuses, such as the John Deere Headquarters, in which practically every tree and landscape element is meticulously placed and clearly designated. In the Weyerhaeuser plan the forest is rendered, or rather, unrendered, as white terra incognita, and planted trees marked on the plan simply give it frontage. The plan, in other words, was marking a beginning: "the site needed more repair than design," says Walker, "so we worked with a forester from Weyerhaeuser, and we learned how to build a forest."[61]

And so, the process of engineering the forest began, somewhat akin to what Weyerhaeuser was doing in its tree farms. The company's foresters worked on establishing the planting rationale, aiming at a forest that could last over the long term. Due to its planning, visibility, and accessibility, the site quickly established itself as a destination for visitors and hikers from the surrounding region. Weyerhaeuser, conscious of the inherent public relations benefits in direct experience, did not stop them, up to the point where people were allegedly walking the covered pathways adjacent to the working areas on their way through the site.[62] As more pieces were added to the campus, the forest was understood by the architects and landscape architects as the matrix holding it all together, and deeply affecting the design and experience of both employees and visitors. The forest on the Weyerhaeuser campus was a real setting with its own set of characteristics, but at the same time, it presented the engineered forest metaphor as a working model through which specific types of architectural approaches and specific types of buildings, landscapes, and interiors could be developed.[63] The unusual opening up of the Weyerhaeuser campus and its growing popularity as a recreational destination led to the view that it was a conscious response by the company to Earth Day and the rise of the environmental movement, in which "many corporate estates explicitly used their landscape spaces to orchestrate an environmentally positive image of their businesses".[64] This may be an oversimplification, as the building and surrounding landscapes were largely finished by Earth Day, but George Weyerhaeuser did recognize the impact of the raging sentiments in California and Washington on the attitudes of the company.[65]

There was, however, another role to be played by the project. The company's 1971 report, distributed to shareholders, emphasized the central role of the headquarters building in its business strategy by placing a gleaming night shot, taken by Ezra Stoller, on the cover. Inside, among the flow charts and reassuring words of pecuniary optimism, was another photograph. Taken from the upper floor, it depicted the company's executives seated nonchalantly around a long wooden table covered with paper reports and coffee cups. Behind them, a large glass window overlooks the pond and surrounding forests that ostensibly only begin their regeneration. This photograph can be understood in two ways. In one version, the captains of industry have found refuge from the instabilities of the times in articulating their forest castle as an edifice of the environmental era. In the other, they are scanning their holdings while planning their retaliation. This is because, at the time this photo was taken, Weyerhaeuser was already exploring tropical forests in Brazil and Southeast Asia as potential sources of supply. In addition to its massive 1969 land acquisition of 1.8-million-acre Dierks Forests' lands in Oklahoma and Arkansas, this signaled the beginning of a gradual divestiture from the Pacific Northwest to areas with less regulation and less public attention (Figure 1.10).

Flowing from modified woods meant to increase production for imperial fleets, to the hard rationale of plantations in tropical contexts, through orangeries and terraria that mirror the tastes and means of ruling classes, and up to corporate utilization of forest plantations as idyllic open landscapes in Washington or for creating wooded hills covering a mall in

FIGURE 1.10   Weyerhaeuser executive floor, c. 1971. © Weyerhaeuser.

China, the metaphor of the engineered forest presents the possibility of a construct that is demonstrative of radical reinventions of natural environments. The desire to rationalize such environments also speaks to an underlying theme present in various histories of designed forests, binding intense economic activities and mystified presentations to the public, filtered through the forest green.

Notes

1 Ravenscroft, Tom. "Steven Chilton Architects Completes 'Bamboo Forest' Theatre in Wuxi." December 21, 2019. *Dezeen*. Accessed December 10, 2023. https://www.dezeen.com/2019/12/21/steven-chilton-architects-wuxi-taihu-show-theatre-china/.

2 Griffiths, Alyn. "Colliding Gabled Structures Form Hannae Forest of Wisdom Community Centre." April 28, 2019. *Dezeen*. Accessed December 10, 2023. https://www.dezeen.com/2019/04/28/unsangdong-architects-hannae-forest-wisdom-seoul-community-centre/.

3 Spears, Tom. "MINI LIVING Forests Space by Asif Khan at London Design Festival." September 15, 2016. *Designboom*. Accessed December 10, 2023. https://www.designboom.com/design/mini-living-forests-asif-khan-london-design-festival-09-15-2016/.

4 "White Forest in Monsanto/Bruno Camara Arquitectos." August 20, 2019. *ArchDaily*. Accessed December 10, 2023. https://www.archdaily.com/923245/white-forest-in-monsanto-bruno-camara-arquitectos.

5 Fehrenbacher, Jill. "Air Forest: Inflatable Pavilion Lights up the Night." June 29, 2009. *Inhabitat*. Accessed December 10, 2023. https://inhabitat.com/air-forest-inflatable-outdoor-pavilion/.

6 While the Avery Index covers a wide range of periodical literature from all five continents, it has an inherent bias toward publications in the English language and towards "developed" economies. The empirical sortie is therefore indicative only of mainstream attitudes toward forests in architecture.

7 The publications are *A + U: Architecture and Urbanism, A10: New European Architecture, Ab/Abitare, Architecture, Architecture* (New York), *Architecture* (Washington DC), *Architecture Australia, Architecture d'Aujourd'hui, Architecture Today, Architectural Design, Architectural Digest, Architectural Forum, Architectural Record, Architectural Review, Architectural Review* (Boston), *Architektur, Innenarchitektur, Technischer Ausbau, Arquitectura Viva, Arkitektur* (Sweden), *AV Proyectos, Azure, Bauwelt, Blueprint* (London, England), *Buildings, Build Kenya, Canadian Architect, Casabella, Cite, C3 Korea, Croquis, Detail, Detail (English Ed.), Dwell, GA Document, GA Houses, GA Japan, GA Japan: Environmental Design, Harvard Design Magazine, Inland Architect, Inland Architect and News, JA, Japan Architect, Journal of Architectural Education, Journal of Architecture, Journal of the American Planning Association, Metropolis, Planning, Progressive Architecture, RIBA Journal, Space, Urban China, Urban Land, Wallpaper, Western Architect, Werk, Bauen + Wohnen*.

8 "Alvar Aalto in the Finnish Forests, the Virtuoso of Word Puts up a Civic Center of Brick." *Architectural Forum* 100 (April 1, 1954): 148–51.

9 "Forest Murmurs." *Architectural Review* 157 (June 1975): 386.

10 Manroe, Candace. "Pine Forest Palette: Rustic Storybook Cabin High in the Rocky Mountains." *Architectural Digest* 41, no. 6 (1984): 66. Mays, Vernon. "A Machine in the Forest: Center for Innovative Technology, Herndon, VA." *Progressive Architecture* 70, no. 8 (1989): 92.

11 Pollock, Naomi R. "A Forest Runs through It." *Architectural Record* 183, no. 7 (1995): 34–7.

12 Woodman, Ellis. "Paradise Lost." *Architectural Review* 237, no. 1416 (2015): 44–55. Cunha, Dilip da. "The Jungle's Call." *Harvard Design Magazine*, no. 45 (March 2018): 228–31.

13 "1000 Trees/Heatherwick Studio." January 18, 2022. *ArchDaily*. Accessed December 11, 2023. https://www.archdaily.com/975297/1000-trees-heatherwick-studio.

14 Janick, Jules, and Harry Paris. "History of Controlled Environment Horticulture: Ancient Origins", *HortScience* 57, no. 2 (2022): 236–8, accessed November 19, 2023.

15 Agricultural societies around the world had been experimenting with controlled environments for a long time, including in China, where the Hanfang system, protecting crops with bamboo and rice paper, is described in the sixth-century agricultural text *Qimin Yaoshu*. Archeological evidence also suggests the possibility of covered crops in ancient Egypt and Mesopotamia, among other regions.

16 Around 1760, good-quality cast iron became readily available, allowing for more imaginative uses of the material in construction. Glass roofs were rare before 1800 but were gradually introduced in greenhouses as new varieties of exotic plants that required climatic control all year round were brought into gardens in Europe. See Woods, May, and Arete Warren. *Glass Houses: A History of Greenhouses, Orangeries and Conservatories*. New York: Rizzoli, 1988, 88–141.

17 Plant science advanced dramatically in the first decades of the twentieth century, making the work of these research institutes applicable and at times profitable. Research uncovered both the inner workings of plant physiology, discovering the workings of hormones and nutrients and how plants respond to their environment through biochemical processes.

18 Banham, Reyner, "A Home Is Not a House", *Art in America*, 53, no. 2 (April 1965): 70–9.

19 Banham, Reyner. *The Architecture of the Well-Tempered Environment*. Chicago: University of Chicago Press, 1969.

20 The latter category includes many of Buckminster Fuller's works, including the Dome over Manhattan (1959) and the Biosphere for Expo 67 in Montreal (1967), which made a splash in architectural circles.

21 "Una Grande Serra per Uffici a New York = The New Headquarters for the Ford Foundation in New York." *Domus*, no. 462 (May 1968): 1–6.

22 Gissen, David. *Manhattan Atmospheres: Architecture, the Interior Environment, and Urban Crisis*. Minneapolis: University of Minnesota Press, 2014.

23 Roche, Kevin, John W. Cook, and Heinrich Klotz. "Kevin Roche Interview." *Perspecta* 40 (2008): 104–19.

24 Ada Louise Huxtable. "The Ford Foundation Flies High." *New York Times*, October 26, 1967, sec. II: 23, 25. Reprinted in Huxtable, *Will They Ever Finish Bruckner Boulevard?* New York: Macmillan, 1970, 90–91.

25 Roche and Dinkeloo continued to explore the idea of radical environmental control in projects such as the extensions and renovations of the Metropolitan Museum of Art (completed between 1974 and 1985), which included climatic-controlled halls, and the Deere & Company West Office building in Moline, Illinois (1979) which featured a "winter garden" in the atrium. See Gissen, David. "The Architectural Production of Nature, Dendur/New York." *Grey Room*, no. 34 (2009): 58–79, and Roche, Kevin, and Francesco Dal Co. *Kevin Roche.* New York: Rizzoli, 1985.

26 I follow David Gissen's term here.

27 Masello, David. "Raymond Jungles Reshapes the Garden at the Ford Foundation Overhaul." *The Architect's Newspaper*, April 8, 2019. https://www.archpaper.com/2019/04/raymond-jungles-reshapes-garden-ford-foundation-overhaul/.

28 Moore, Jason W. "Madeira, Sugar, and the Conquest of Nature in the 'First' Sixteenth Century: Part I: From 'Island of Timber' to Sugar Revolution, 1420–1506." *Review (Fernand Braudel Center)* 32, no. 4 (2009): 350.

29 Evelyn's book *Sylva: Or A Discourse of Forest-Trees and the Propagation of Timber in His Majesty's Dominion* was published in 1664 and recognized up to this day as one of the pioneering works in the field of forestry.

30 Grove, Richard. *Green Imperialism: Colonial Expansion, Tropical Island Edens, and the Origins of Environmentalism, 1600–1860.* Cambridge; New York: Cambridge University Press, 1995.

31 Buffon's ideas will be discussed in the next chapter, in relation to his larger conceptual formulation of natural history.

32 "Sucre." Encyclopédie ou Dictionnaire raisonné des sciences, des arts et des métiers, 15: 608–14 (Paris, 1765).

33 "Sucre." Encyclopédie ou Dictionnaire raisonné des sciences, des arts et des métiers, 15: 609 (Paris, 1765).

34 Jean-Baptiste-Pierre Le Romain, who contributed the entry, was an engineer who lived in Martinique and wrote several articles on the Caribbean. He gained notoriety in recent scholarship for his racist views of indigenous peoples, also included in the Encyclopédie. See "Sucrerie." Encyclopédie ou Dictionnaire raisonné des sciences, des arts et des métiers, 15: 618 (Paris, 1765).

35 In the 1750s, du Monceau began working on a comprehensive multi-volume work titled *Traité complet des Bois et des Forêts*, which included two volumes titled *Traité des Arbres et Arbustes* (1755), *La Physique des Arbres* (1758), *Des semis et plantations des arbres* (1760), *De l'Exploitation des Bois* (1764), and *Du Transport* (1767).

36 Du Monceau, *Des semis et plantations des arbres et de leur culture*, 352.

37 Olmsted, Frederick Law, and Charles E. Beveridge. *Writings on Landscape, Culture, and Society.* New York, N.Y.: Library of America, 2015, 737.

38 Olmsted showed earlier interest in scientific land management, including forestlands. In 1880, he designed the 275-acre Moraine farm in Beverly, Massachusetts, which included a small 75-acre experimental forest. See Thoren, Roxi. "Deep Roots: Foundations of Forestry in American Landscape Architecture." *SCENARIO 04: Building the Urban Forest*, Spring 2014. https://scenariojournal.com/article/deep-roots/.

39 Olmsted, Frederick Law, and Charles E. Beveridge. *Writings on Landscape, Culture, and Society*. New York, N.Y.: Library of America, 2015, 661.

40 Olmsted meticulously describes his forest management strategy:

> remove all the old oaks that are not of exceptional and admirable character; let the thickets of the younger oaks be judiciously thinned; give the other trees (Hickories, Chestnuts, Limes, (Basswood) Tupelos, Beeches, Maples, Tulips, Birches,) that are sparsely growing with them, a fair chance; plant occasional vacant spaces with yet other trees, natural to the circumstances, such as I have named; encourage a growth of underwood, and a forest would result that would easily come in time to be the finest in the country.
>
> *Olmsted and Beveridge, 2015*

41 Olmsted advised Vanderbilt that large portions of the estate should be planted with white pines for aesthetic and sensorial enjoyment, but not less importantly, for future economic value.

42 The process was documented in a publication and presented during the exposition. See Pinchot, Gifford. *Biltmore Forest: The Property of Mr. George W. Vanderbilt; An Account of Its Treatment, and the Results of the First Year's Work*. Chicago: R.R. Donnelley & Sons Co., 1893.

43 Banham, Reyner. *Megastructure: Urban Futures of the Recent Past*. London: Thames and Hudson, 1976, 19.

44 Blake, Peter. "Slaughter on Sixth Avenue." *Architectural Forum* 122, no. 3 (1965): 18.

45 Martin, Reinhold. "Atrocities. Or, Curtain Wall as Mass Medium." *Perspecta* 32 (2001): 70.

46 Canty, Donald. "Evaluation of an Open Space Landscape: Weyerhaeuser Co." *AIA Journal* 66, no. 8 (July 1977): 40. This formal version that went out to the press covered a rocky beginning: Bunshaft was in fact fired after a disastrous presentation and SOM kept the project only through the intervention of Louis Skidmore, who proposed Bassett would lead it. Sasaki, Walker and Associates remained in the project due to George Weyerhauser's interest in its Foothill College project at Los Altos Hills (1960). Handel, Dan. Interview with Peter Walker. [Online interview]. December 6, 2023.

47 The creation of Weyerhaeuser is indicative of a process in the Pacific Northwest in which land was ceded from indigenous tribes to the United States in the 1850s and then transferred from the government to railroad companies through land grants and then to private landowners associated with these companies,

who sold off the land for profit. While the Great Northern is often regarded as a company that did not rely heavily on land grants, its economic history shows that existing subsidies in fact enabled Hill's initial purchase of the St. Paul and Pacific Railroad company and that the land acquired on the way to Puget Sound was a result of a deal with the government that was based on these previous grants. See Rae, John B. "The Great Northern's Land Grant." *The Journal of Economic History* 12, no. 2 (1952): 140–45.

48 In 1905, Washington became the nation's leading producer of timber, a position it held until the late 1930s.

49 Sharp, Paul F. "The Tree Farm Movement: Its Origin and Development." *Agricultural History* 23, no. 1 (1949): 41.

50 Emily K. Brock. "Tree Farms on Display: Presenting Industrial Forests to the Public in the Pacific Northwest, 1941–1960." *Oregon Historical Quarterly* 113, no. 4 (2012): 537.

51 Emily K. Brock. "Tree Farms on Display: Presenting Industrial Forests to the Public in the Pacific Northwest, 1941–1960." *Oregon Historical Quarterly* 113, no. 4 (2012): 526.

52 Brock follows the extensive public relations campaign by industry, which included child-centered publications and activities, based on the idea of investment in the next generation of voting citizens. Emily K. Brock. "Tree Farms on Display: Presenting Industrial Forests to the Public in the Pacific Northwest, 1941–1960." *Oregon Historical Quarterly* 113, no. 4 (2012): 542.

53 "When we were looking at one of the sites, walking in a forest I said, 'this is really beautiful' and George said 'you obviously don't know anything because these trees are the first ones we would be cutting, since they're worthless.'" Interview with Peter Walker. [Online interview]. December 6, 2023.

54 Charles Bassett, "Oral History of Edward Charles Bassett" interviewed by Betty J. Blum, compiled under the auspices of the Chicago Architects Oral History Project, The Ernest R. Graham Study Center for Architectural Drawings, Department of Architecture, The Art Institute of Chicago, 1992, 105.

55 Charles Bassett, "Oral History of Edward Charles Bassett" interviewed by Betty J. Blum. Walker described the scene in retrospect:

> Chuck was really a teacher. He had a group of young people working on this [project]. At the time, the office was all white and you could pin up on the white walls, and he [Bassett] had no desk where he sat, only a drafting table, so it was all very flexible. What he asked us all to do was to look at the site and find the possibilities where the building could be. And we thought we would put it up on the hills, and then Chuck went up and drew it with his finger, saying 'what if it would be between these two hills?'.
> *Interview with Peter Walker. [Online interview]. December 6, 2023*

56 Interview with Peter Walker. [Online interview]. December 6, 2023.

57 Interview with Peter Walker. [Online interview]. December 6, 2023.

58 Bassett, "Oral History of Edward Charles Bassett", 105.

59 Montgomery, Roger. "A Building That Makes Its Own Landscape." *Architectural Forum* 136, no. 2 (1972): 20.

60 Bassett remarked that "The climate in the Northwest is so beautiful with the mists and fogs and the grayed greens. Very Whistler-like, very oriental." Bassett, "Oral History of Edward Charles Bassett", 112.

61 Interview with Peter Walker. [Online interview]. December 6, 2023.

62 Dave Dickerson, facility manager for the campus, said,

> I don't think [public use] was actually encouraged, but it was never discouraged. People were always welcome there. We appreciated people. We had a lot of conversations with people in the area that would walk on [the campus] and about how much they enjoyed it.
>
> *Dickerson, Dave. Interview by Spencer Howard.*
> *Northwest Vernacular, December 14, 2020*

Transcription prepared by Jean Parietti. Save the Weyerhaeuser Campus. Project file 2020_019 Weyerhaeuser Campus, Seattle, WA.

63 The interior design of the project, developed by SOM and Knoll, was saturated with forest metaphors. It both included visual elements such as tapestries or living plants that layered the experience of being in and looking out of the building, and presented an innovative organizational approach, drawn from European *Bürolandschaft* schemes, in which the internal organization of workstations and other spatial elements would frequently shift based on the needs of the corporation. This would sometimes result in a layout that changed every three weeks. See Handel, Dan. "Landscapes of Co-Option Soft Power and the Environmental Turn in Corporate America." In *Bracket 2: Goes Soft*, edited by Neeraj Bhatia and Lola Shepard, 141–4. Barcelona/New York: Actar, 2013.

64 Mozingo, Louise. *Pastoral Capitalism: A History of Suburban Corporate Landscapes.* Cambridge, MA: The MIT Press, 2011, 140. Mozingo writes:

> The first corporate estate on the West Coast, and still the only one that rivals in scale and grandeur its East and Midwest counterparts, the 1971 Weyerhaeuser Corporate Headquarters outside Tacoma, Washington, further articulated the corporate estate as a vehicle for public display.

65 At the same time, Weyerhaeuser underplayed the effects of environmental discourse and legislation on the company's business. Weyerhaeuser, George H. "Oral History," interviewed by Linda Edgerly, compiled under the Forest History Society Oral History Collection, October 10, 1986, p3/4042/08b-277.

2

JUNGLE

In 1955, Mies van der Rohe, the famed architect who directed both the Bauhaus and the Illinois Institute of Technology, an icon of the modernist conquest of reality, reflected that

> [t]here are no cities anymore ... it goes on like a forest ... that is the reason why we cannot have the old ... planned city and so on. We should think about the means that we have to live in a jungle, and maybe we do well by that.[1]

Judging by the tone and content of this statement, one might have thought that Mies was sharing a hut with Martin Heidegger in the Black Forest, but he was, in fact, well settled in his Chicago base, where forests could not be seen for hundreds of miles around. The curious use of forest metaphors was reasoned by eminent Mies scholar Detlef Mertins as part of the architect's ongoing quest to define organic forms in both architecture and urban design. "By the mid-1950s," he observes, "Mies had reconciled himself to the disappearance of the city ... without center or edge, this new territorial urbanism was transforming the crust of the earth into a hybrid of nature and technology."[2] His call to "maybe do well by that" was therefore simply a pragmatic redefinition of the roles of architecture in a sprawling urban landscape. However, anyone else would be hard-pressed to find an optimist call to action in this statement. In fact, the disappearance of planned cities is told in a tragic tone, as if presented by Oswald Spengler: the drowning of human attempts to introduce order is a done deal, and "we" can only learn to live with it. The forest that "goes on" and the jungle that surrounds architects and urban designers are a far cry from the organic metaphors of harmonious environments that support the growth of true form. They are hostile, chaotic, and uncivilized areas that blatantly resist the core concepts of architectural modernism.

                    DOI: 10.4324/9781003473411-3

If Mies's eulogy had to come with an illustration, it could be found in the contemporary work of his colleague and fellow émigré Ludwig Hilberseimer. Appreciated and abhorred by generations of architects due to his radically dreary urban visions, Hilberseimer took on the task of ceaselessly repeating the point of creating the most disheartening outlooks for the future of cities. Hilberseimer followed Mies to Chicago in 1938 and began working at the Illinois Institute of Technology (IIT), where he embarked on a new cycle of work. Mertins writes that Mies supported regional planning at IIT and Hilberseimer "in developing yet another model of settlement that would [be] … capable of imparting order to the seemingly wild phenomena of urban growth and form in America."[3] Hilberseimer's speculative work was in tune with the sweeping decentralization of urban settlements, inspired by the paranoia of Cold War politics. This was a time when both presidents and evangelists advocated for an escape from cities – always corrupt, deteriorating, and risky in the American imagination – beyond the radiation zone to the safety of a suburban enclave. Hilberseimer's books were meant to complement this sentiment and give new relevance to his previous large-scale visions. Ironically, the success of these visions, as Joseph Rykwert noted, was in their prophesying "a city without streets, even worse, if that is possible, than that which has actually been realized."[4] Hilberseimer based his coordinated assault on American space on the Settlement Unit. This was essentially a suburban island contrived as a telephone pole figure of roads, seamlessly connected to a superhighway and so to the next identical unit, *ad infinitum*. In the plans of the Settlement Unit, trees are featured as a means to mitigate the rigid geometry of obsessive rationalism. These were not simply trees: the resurgence of plant life was also meant as an act of camouflage. "The city", Hilberseimer said, "will be within the landscape, and the landscape within the city." This idea departed from architectural modernism's ideas of a tower in the park, or a green belt masking the grungy aesthetics of a nearby plant, to present an endless city that was designed to disappear. In the Atomic Age, this proposition made perfect sense. But even hypothetical trees can refuse a supporting role. In the perspective drawings that accompanied the plans, taken from a low aerial angle, they become forest-like, taking over the scene entirely. What the architecture is becomes impossible to tell as an organic softness – absent in the plans – is rendered through thousands of ink dots and moves to the fore. Alfred Caldwell, who drew these perspectives, was an instrumental link between the two older *herren* and the Midwestern cultural scene.[5] His skilled hand subversively translated the urbanist's work, lending it a sensitivity that was lacking in the original, and portraying America's future cities as taken by a jungle (Figure 2.1).

★★★

But perhaps the jungle that came back to haunt the imagination of the two architects had to do with views of primaveral forests that circulated in Wilhelmine Germany as they were making their first professional steps. During that time, the imperial ambitions of the nation found in South America a reflection of its fantasies.[6] The late nineteenth-century German immigration, especially to Brazil, was accompanied by promotional literature that focused heavily on the *urwald* as a site for the shaping of character. Earlier depictions by travelers and scientists, most famously by Alexander von Humboldt, which highlighted the overwhelming beauty and richness of life in tropical forests, were gradually overshadowed by accounts of the primeval forest as dangerous, chaotic, and inherently hostile. Such descriptions made implicit connections between the biological life of the *urwald*, described in negative terms, and its harmful mental and cultural effects. German writings on South America were heavily influenced by Henry Morton Stanley's accounts of African rainforests, translated in the late 1870s and widely circulated, which portrayed his journeys in the light of constant struggle against nature.[7] The cultivation of real forests in Brazil was often presented in Germany in the terms of an armed conflict, which subjected the forest to tillage and to the colonist's eventual prosperity. As indeed it was: the arrival of German colonizers to the "virgin" forest frontiers often followed the violent eviction of the indigenous populations who inhabited these regions.[8] On other occasions, settlers were directly involved in the eradication of entire tribes.[9] The immigrants, in line with voices in the Brazilian federal states, saw this conflict as an inevitable outcome in the march of progress. The jungle in the late nineteenth and early twentieth century was thus associated with conflict, violence, and the need for clearance from both indigenous plants and humans (Figure 2.2).

The negative space the jungle inhabits in the colonial imagination is sometimes described as an outcome of notions of social Darwinism, which became prevalent at the time.[10] These ideas, focused on the struggle between species, supposedly overturned earlier Edenic descriptions of harmonious forest environments. But disparaging accounts of faraway forests had earlier origins in both scientific and travel literature that was, at least until the middle of the eighteenth century, often indistinguishable. Mary Louise Pratt, describing the writings produced following what

45

FIGURE 2.1 Landscape perspective for the Settlement Unit, Alfred Caldwell and Ludwig Hilberseimer, 1944. DR1983:1099 Collection Canadian Centre for Architecture. Gift of Alfred Caldwell.

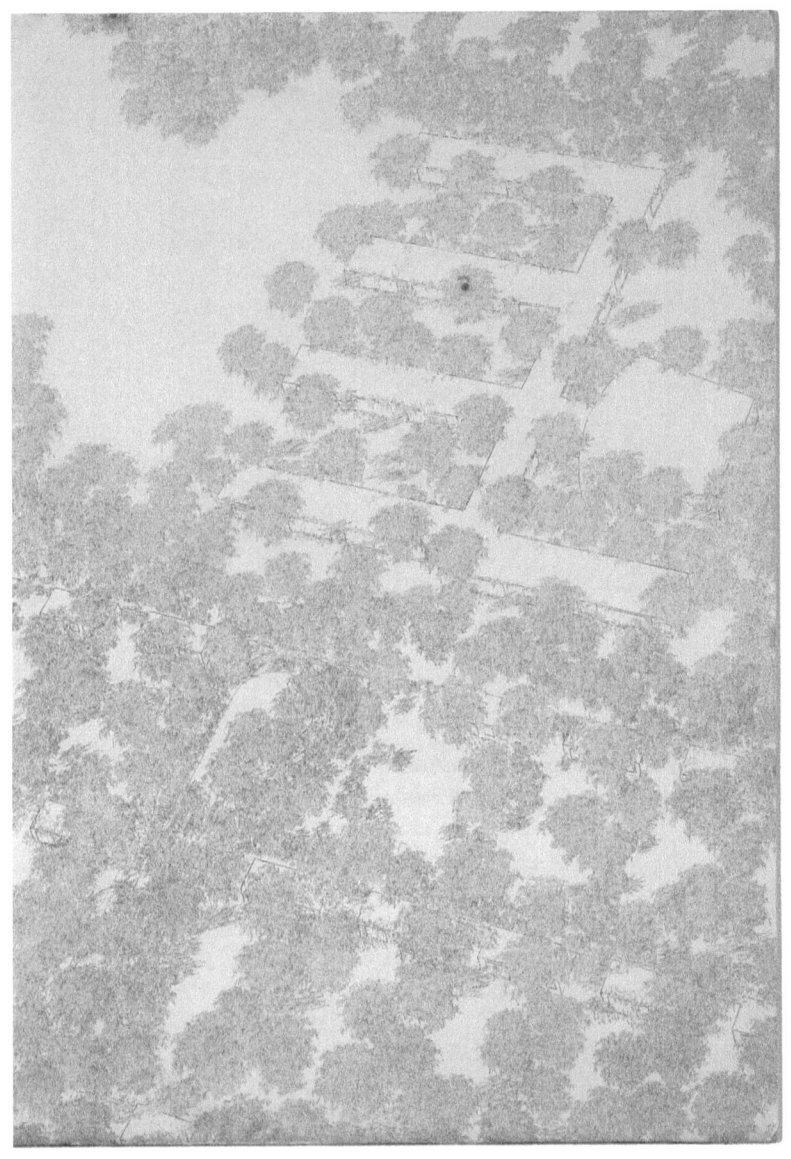

FIGURE 2.2   Virgin forest in Brazil, Comte de Clarac, circa 1820. © Creative
Commons. Painting in the Acervo da Pinacoteca do Estado de
São Paulo, Brazil.

is often described as the first major scientific expedition – the Spanish
French Geodesic Mission to Ecuador – suggests that

> the body of texts that resulted from the La Condamine expedition
> [as it became known] suggests rather well the range and variety of
> writing produced by travel in the mid-eighteenth century, writing
> that in turn produced other parts of the world for the imaginations
> of Europeans.[11]

La Condemine's 1745 report to the Academy of Sciences, and the following journal of his voyage down the Amazon, published in 1751, made
him famous in scientific and intellectual circles. Pratt notes that his prose
is more akin to survival literature than to scientific writing, placing him
as he enters the jungle "in a new world, far from all human commerce,
upon a sea of fresh water…. I met there with new plants, new animals,
and new men."[12]

Soon after his return, La Condamine donated the specimen and art
objects collected in his journeys to Comte de Buffon at the Jardin de
Roi. Buffon, mentioned in the previous chapter for his affiliation with

Duhamel du Monceau, was the most influential European naturalist until Humboldt. It was under his purview that the idea of tropical forests was transformed and loaded with new meanings. Beginning in 1739, Buffon published his *Natural History*, a monumental work in thirty-six volumes that offered an exhaustive compendium of every animal known to exist around the world. This work was based on his access to the Jardin's collections and, later, on countless specimens he acquired from the four corners of the earth. The transportation of dead animals *en masse* substituted field study, and Buffon did little to deny his reputation as never observing the animals in their living state, quite an astonishing fact when one reads his theories on the environmental impacts on species development.

The same arrogance was directed towards the natural world of the Americas. In the twelve-volume section of his sweeping history dedicated to quadrupeds, Buffon weaved a tale of a continent whose climate shaped an entire ecology of losers. Fewer species existed in the New World, he argued, and in cases when the same animal was to be found on both sides of the Atlantic, it would be smaller, feebler on the American side. Buffon concludes that human populations, in this case the native peoples of the land, were in themselves derelict, shy, morally undeveloped, and simple-minded. This claim had more to do with his general ideas about nature than with racial classification, which had not yet developed in European contexts.[13] Basing his hypothesis on accounts by travelers who were themselves plagued by preconceptions and misunderstandings, he described the New World as a cold and damp world, where forests collaborated with swamps to create conditions in which everything proves abortive:

> the earth being every where covered with trees and coarse weeds, it never dries, but constantly produces humid and unwholesome exhalations. In these gloomy regions, Nature remains concealed under her old garments … totally neglected, her productions languish, become corrupted, and are prematurely destroyed.[14]

In that, he was not first: earlier Europeans looked down upon the New World from the day of its discovery: responding to the observation that trees in Hispaniola developed shallow roots, Queen Isabella of Spain wrote that "this land, where the trees are not firmly rooted, must produce men of little truthfulness and less constancy."[15] But here was, for a first time, a treatise that aligned these sentiments with a consistent, rational worldview, buttressed by all the available methods of contemporary Enlightenment science. The animosity with which Buffon portrayed the ecologies of the places he never visited was expressing the anxieties of his time, spurred by the French inferiority in the imperial race and by an impulse to reassert the dominance of European culture.[16] But they were nevertheless instrumental in shaping subsequent encounters with the tropics.

★★★

Later generations of naturalists, many of them influenced by Humboldt's focus on empirical data, searched for ways to reconcile the cultural metaphors of chaotic jungles with contemporary ways of doing science. In the mid-nineteenth century, as scientific forestry was gaining traction in European countries, its encounters with the realities of the Indian subcontinent circumscribed the subfield of tropical forestry. Chaos is sometimes just another name for complexity, and tropical forests threw naturalists off balance partly because once studied, they presented an overwhelming abundance of plant species. Whereas the number of known indigenous trees in nineteenth-century Europe was 158, Indian forests contained 1,200 species, and many known plants took different forms in local climatic and ecological conditions.[17] Dietrich Brandis, the German forester who specified these numbers, was the quintessential figure of the international forest expert, basing his estimations and predictions on data and rigorous methodologies. Yet, the origin myth of tropical forestry, in which he played a central role, carried many of the motifs of survivalist jungle encounters:

> Brandis, so he told me, had traversed the woods of Pegu riding an elephant on such trails as there were, with four sticks in his left hand and a pocketknife in his right. Whenever he saw in the bamboo thickets a teak tree within two hundred feet of his trail, he cut a notch in stick number 1, 2, 3, or 4, denoting the diameter of the tree. It was impossible for European hands, dripping with moisture, to carry a notebook. At the end of the day, after traveling some twenty miles, Brandis had collected forest stand data for a sample plot four hundred feet wide and twenty miles long, containing some nineteen hundred acres. He continued his cruise for a number of months, sick with malaria in a hellish climate. Moreover, he underwent a trepanning operation, and for the rest of his life he carried a small hole filled with white cotton in the front of his skull. But he emerged from the cruise with the knowledge needed for his great enterprise.[18]

This recount described Brandis's work in colonial India as stemming from a heroic survey mission of a lone forester, in which new findings informed new conclusions; these, in turn, enshrined new principles for

FIGURE 2.3   Self-portrait by Dietrich Brandis at his forest field camp, Burma, circa 1857.

forest growth management in tropical regions and large-scale plans that transformed forests around the world (Figure 2.3).

Brandis is widely considered the most immanent professional forester of his time, and much of his fame as both scientist and administrator resulted from his long career in British India. In terms of the sheer size, the number of people involved, and the volume of economic activity, the management of Indian forests was by far the most influential model for over a century and a clear demonstration of the power of foresters to command and transform entire continents. However, tropical forestry was far from an immaculate theory arising in full form during that *eureka* moment in Pegu. It was rather a gradually evolving, complex system of environmental transformation, involving proved botanical principles and untested predispositions, often shaped on the run within a flux of colonialist pressures and intercultural conflicts.

Born in 1824, Brandis was educated in botany and plant chemistry in Copenhagen, Göttingen, and Bonn, where he took on a position as lecturer in 1849. This somewhat standard career path took a turn after he married Rachel Marshman, an Englishwoman brought up in India.[19] It was probably through Rachel's brother, the journalist John Clark

Marshman, that Brandis was offered by Governor-General of India, Lord Dalhousie, the position of Superintendent of Teak Forests in the Pegu region of Burma, which at the time was the eastern province of British India. While Brandis was initially more interested in a botanical excursion to the region, he decided to accept the job. Gathering scientific equipment and a letter of recommendation from the eighty-six-year-old Humboldt, he left for India in 1855.[20]

Dalhousie's offer can be understood in the context of his attempts to "modernize" India and bring it under government authority. His tenure marked the transition of rulership from the British East India Company to the Raj, in which the Crown assumed control over British territories in the subcontinent.[21] Dalhousie cultivated the image of an enlightened colonial administrator, investing in large infrastructural works and in the creation of the public agencies needed for their realization. At the same time, answering to the company's court of directors, he orchestrated aggressive annexation campaigns known as the doctrine of lapsation, meant to increase the company's territorial holdings and revenue stream, usually under false pretenses. Pegu was the most recent of these acquisitions, even though, being an independent kingdom and not one of the princely states of India, its annexation was technically different. At any case, it stretched the boundary of British rule to the Far East and forced Dalhousie to consider new ways to finance his expansionist ambitions. Recruiting Brandis was the result of his growing conviction that proper forest management would be vital in this effort. Through his communication with Joseph Dalton Hooker, the son of William Hooker, director of the Royal Botanical Gardens in Kew, Dalhousie was informed about the environmental impacts of large-scale deforestation that were then beginning to be recognized. As a result, he felt strongly that there was a need to establish a centralized effort on behalf of the government that would "give this country the clothing of forest trees".[22] Richard Grove wrote in detail about Dalhousie's ambition to create a working forest system in the Punjab and its links to an earlier British experiment in Sind, based in turn on precolonial *shikargahs* – game and forest preserves put in place by the amirs of Sind. Notably, Sir Charles Napier, who annexed the Sind in 1843, came across the *shikargah* system as the forest preserves, often described by British officials as "jungles", were frequently used by Sind guerilla forces.[23] By the time Brandis began his mission in Pegu, the rationale and some of the infrastructure for large-scale colonial forest system were already in place. What Brandis brought to the table was a new level of methodological rigor.

His initial survey of the Pegu forests, taken in the rough conditions described above, was coolly described in his first report, which specified the statistical way his numbers were arrived at, based on the arduous traversing of five regions where teak trees of first (six feet in girth and above) and second class (four and a half to six feet in girth) were counted.[24] Already at this early stage, Brandis acknowledges the value of local expertise, updating the tree categories in accordance with the habits of Burmese "natives" who assisted him in enumeration.[25] The result was an estimate of 584,960 first-class trees, more than twelve times the number that appeared in preceding reports. While the science of tropical trees was still unsubstantiated, Brandis swiftly moved forward with devising a system that would both yield short-term income and begin a process of proper long-term management.[26] The felling process would require the killing of mature trees by girdling and then their seasoning and eventual removal. Achieving the latter objective was more complicated, as Burmese forests presented unique challenges that were unfamiliar in Europe, and even in the young teak forests of Bombay. The mixed nature of the forests, where teak is surrounded by a great number of other species with little value, which often grow much faster, required a change in strategy, and the main task became the consolidation of forest territories and their planned transformation into pure teak forests. Brandis highlights some examples of such forests in precolonial forest reserves, where "[t]he King of Burmah, and several governors ... had the wisdom to declare certain forest districts in their country reserved districts, and forbid the felling of trees."[27] While small in scale, these examples, which often included active planting, set the ideal for what Brandis had in mind for the future.

While this may sound culturally sensitive, Brandis displays at this point in time a rigid approach toward indigenous ways of working the land. Discussing the local Toungya cultivation system, in which teak forests were cut and burned to clear land for crops, he renders it "the greatest enemy of teak in this province" and moves forward with recommending strict measures to prevent it from occurring in Burma's forests. Worse still, the Toungya allowed the jungle to creep back in and take control over the forest. Brandis quotes from an earlier report to the commissioner of the Government of Mysore, which describes the process:

> In these clearings, the primeval forest, with all its beautiful timber and valuable productions, has given place to a thick scrub of noxious weeds and brambles, containing nothing useful. It may be supposed that clearing the forest would make the country more healthy, and so it would if the clearings were more permanent; but the forest is now destroyed only to be replaced by a thick jungle of rank vegetation, still more unhealthy than the forest, which being open below, admits of circulation of air, but the scrub is a dense mass of vegetation, and from bottom to top it is about twenty feet high.[28]

The quote complements the more reserved language of the Brandis report in describing the various injurious effects of the jungle: it is harmful to the health of both (Western) people and (alpha) trees, harbors the dangers of fire and hostile indigenous land practices, and, worse still, is useless from an economic standpoint. And so, the task of tropical forestry, even at this embryonic phase, became one of distilling the forest from the jungle.

Writing about the complex formation of forest acts in colonial India, Kalyanakrishnan Sivaramakrishnan poses that "[t]he history of ecological imperialism would be manifest in the pattern of tree species exploited, planted, and regulated by law and silvicultural science."[29] While promoted under the mantle of objective science and through utilitarian language, the project of tropical forestry was imbued with economic, cultural, and political demands. Brandis, who emerged as "the hero of Pegu" to be appointed the first Inspector General of Forests in India, had a significant role in the first version of the Forest Act. The act, which was passed in 1865, was meant to regulate the supply of timber for railway projects and served as a precedent for the 1878 India Forest Act, which is still largely in force. As Sivaramakrishnan shows, the final version of the act represented a coercive approach towards the indigenous right of cultivation, leading to "unilateral forfeiture" of traditional subsistence.[30] The practice of denying local populations of rights and usage lest they prove ownership in western standards was repeated throughout the colonial world, in Java, Tanganyika, and Indochina.[31]

It is curious that as colonial measures were becoming stricter, Brandis became known also for relaxing his approach toward indigenous cultivation. Already in 1856, a local Karen presented to him a method which combined traditional cultivation with teak planting, providing the saplings protection in their early years. Brandis noted in his report that if the people can be brought to adopt such a system, it may become the most efficient way of afforesting the country with teak.[32] The Toungya system he would later devise, which gave incentives to cultivators for planting, became widespread in India and the colonies and secured his position as an early pioneer of agro-forestry. As he traversed the provinces in his role of Inspector General, he found more and more evidence that precolonial systems of preservation and utilization of forests could be integrated into modern practices of forestry.[33] In his "*sociology* of forest management", writes Ramachandra Guha, "[his] views must be immediately distinguished from almost all other forest officials, Indian or European, before or since."[34]

Beginning in Pegu, Brandis made progress in what he called "working plans". These were documents meant not only to survey and chart existing conditions but to also sketch out the trajectories for achieving forest goals.[35] At times, these plans were provided with detailed maps, and so, the objectives of tropical forestry were given form in a number of large-scale regional planning documents, which presented immense, long-term, and designed transformation of landscapes. The plan for the Pegu and Tenasserim divisions in British Burma is striking in this regard. Compiled in 1881 by Babu Jadanath Nath at the Dehradun forestry school, the sketch map is a culmination of various cartographical efforts in the region. It is based on an 1862 draft by Brandis and an 1881 draft by the Conservators of Forests in the region. In the eastern Attaran province, it is somewhat based on an 1852 forest map made under the supervision of Thomas Latter soon after the British seized Lower Burma.[36] The differences between the two maps speak volumes about a radical change of conception occurring in these decades. From a technical standpoint, the 1852 map is a fine example of contemporary mapmaking, providing a detailed account of the topography and hydrology of the covered area, and a sectional analysis that articulates the land strata and the appearance of teak on clay-rich soils, as opposed to the higher plateaus in which fir and Engaben trees dominate the landscape.[37] The map goes into considerable detail in marking wooded tracts along the Thoungyeen and Attarim rivers and their tributaries, noting specific characteristics such as "wales exhausted" or "not much teak" and marking the names of forest permit holders, all of which are Anglos. The second map disregards many of the features of the land in favor of an unequivocal forest perspective. The rising hills to the east are not even shown, only mentioned in notes, and the network of streams and rivulets are given hierarchy in relation to their capacity of moving drift timber. More significantly, the map is not an image of existing conditions as it is a blend of present and future organization. This is clear as it gives little detail on surrounding areas, which were at the time largely covered in trees, while focusing its intervention on a large forest zone between the Rangoon–Prome railway to the west and the Sittang river to the east.[38] As the map's title suggests, its ambition is to consolidate "sanctioned and proposed" forest reserves into an intensive teak producing area that could be managed over a long time. The technicalities of extracting and shipping the timber via rail and water are therefore marked as part of the plan, highlighting its potential diffusion into local and international markets. The piecing together of the large forest area – approximately 180 miles in length and reaching 25 miles width in certain areas – is procured by fusing together present and future forest reserves with large areas of Karen lands that are left to the locals to cultivate under the Toungya system. Thus the map demonstrates far more than technical or scientific progress being made over a few decades in the latter half of the nineteenth century: it attests the growth of a nuanced, site-specific approach to tropical forestry, which emphasizes local peoples and traditions as part of the management strategy (Figures 2.4 and 2.5).[39]

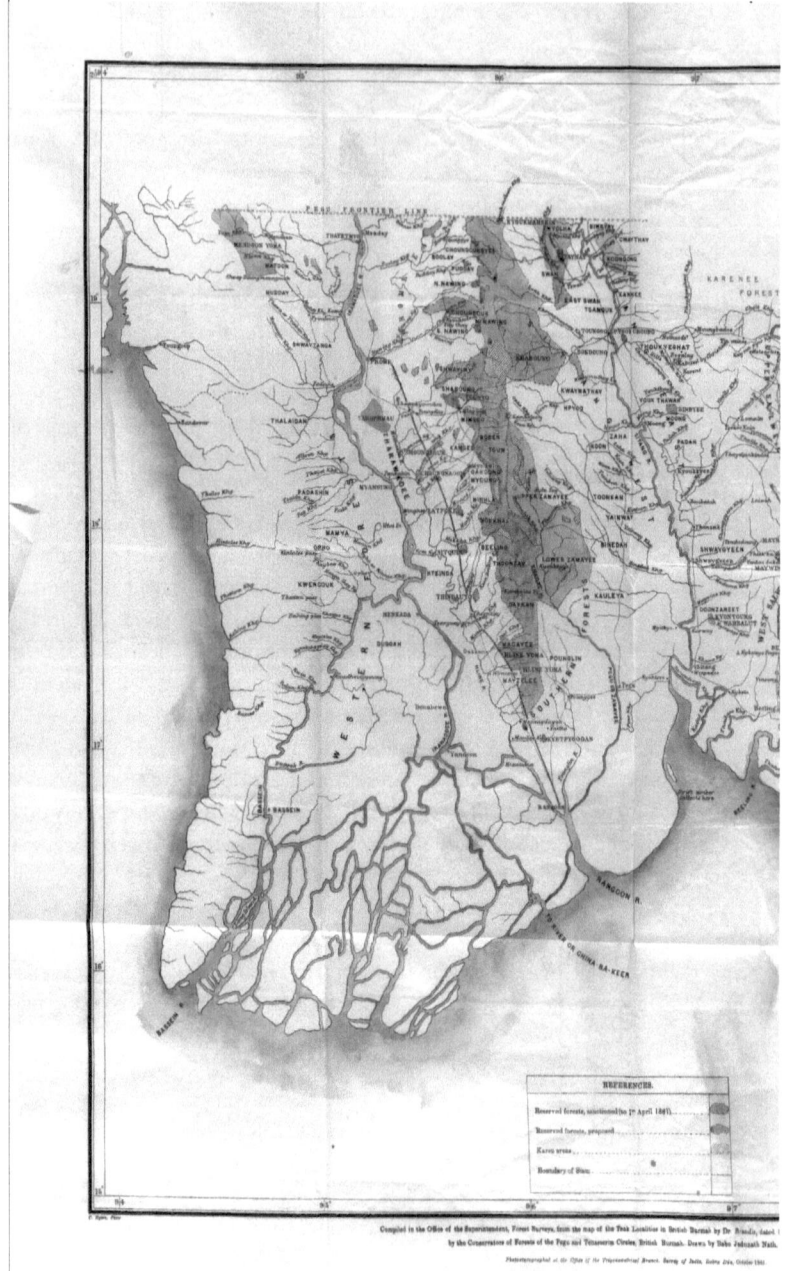

FIGURE 2.4   Forest plan for the Pegu and Tenasserim divisions, 1881. Courtesy
Botany Libraries, Harvard University Herbaria.

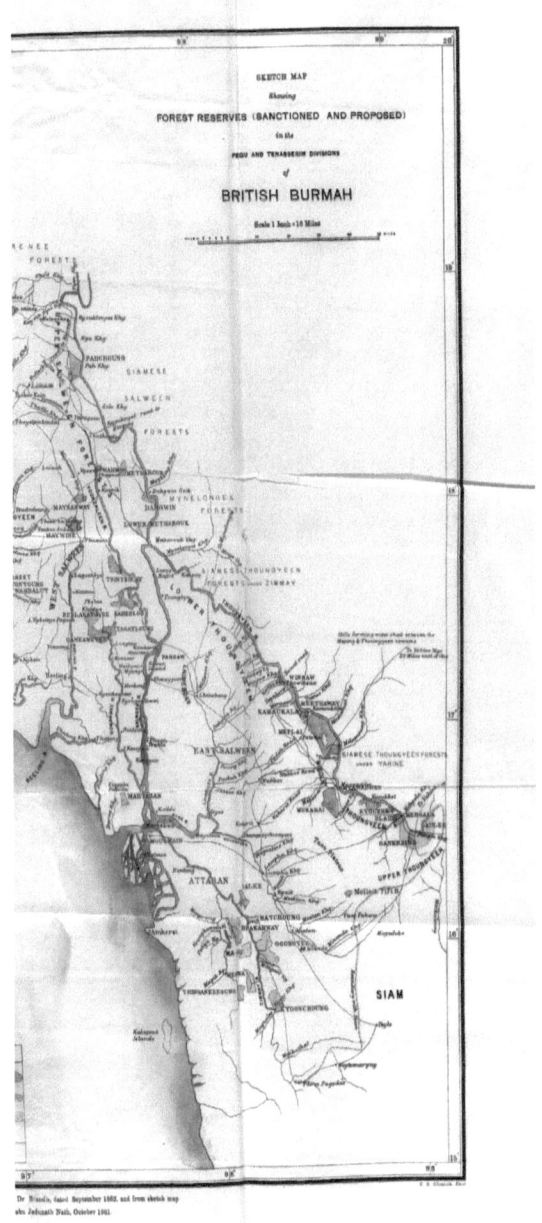

SKETCH MAP
Showing
FOREST RESERVES (SANCTIONED AND PROPOSED)
in the
PEGU AND TENASSERIM DIVISIONS
of

BRITISH BURMAH

Scale 1 Inch = 16 Miles

SIAM

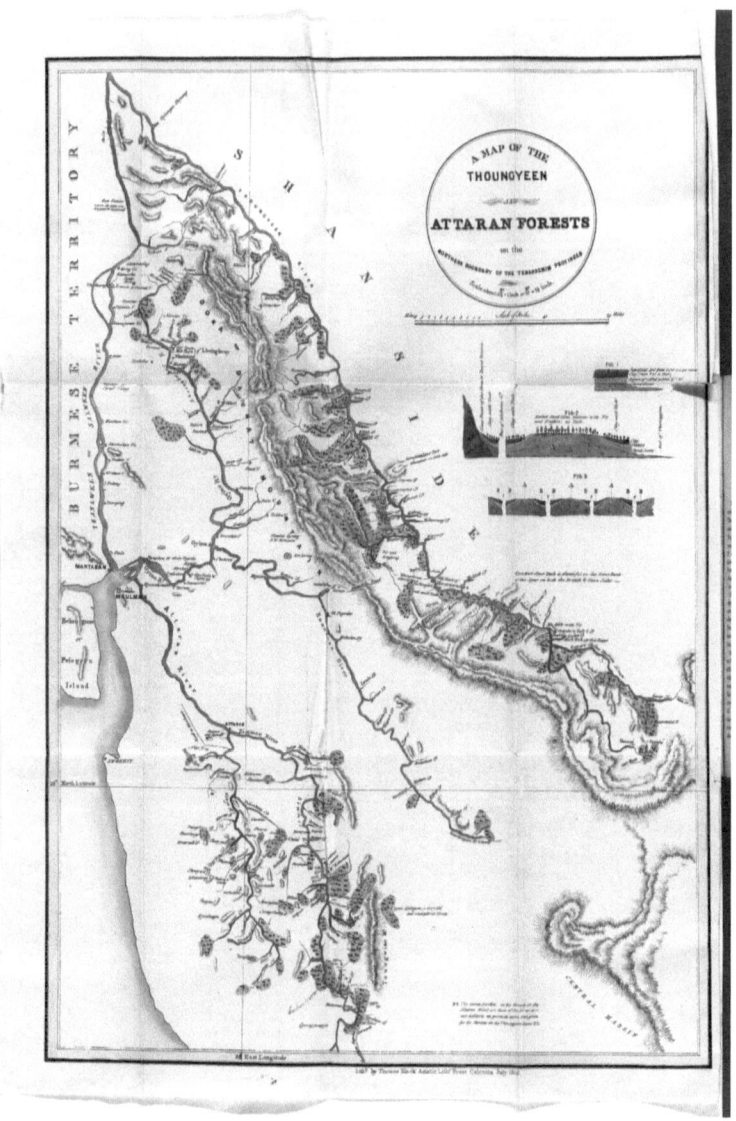

FIGURE 2.5   Forest map of Thoungyeen and Attaran Forests, 1852. Courtesy
Botany Libraries, Harvard University Herbaria.

Considered as design scheme, the sheer magnitude of the area of the plan makes it one of the early forerunners of large-scale landscape design. It would be tempting for the historian to find equivalents in Frederick Law Olmsted's interest in scientific land management, which he did develop in the 1850s and 1860s. But Olmsted's calls for establishing a nursery for Central Park, or his later shots at experimental forests, remain confined to an area few dozen acres, whereas the first time he rises up to the really large scale of redesigning a forest landscape, and perhaps the first time he considers the links between Brandis's work and his own, is around 1890 at Biltmore.[40] It is well known that Biltmore was the springboard for American forestry, which was much influenced by the Indian model. Some scholars went as far as to argue that Indian forestry became intimately entrenched with Empire, as it swiftly spread to new territories.[41] This narrative may be underplaying the particularities of Indian territories as well as the inherent contradiction between the scientific and the administrative positions held by foresters on the ground. Brandis himself, while being perceived by protegees and peers around the world as most successful in his endeavors, reflected that during his long tenure in India, his work was mostly hindered by the shortsightedness and financial obsessions of colonial administrators. "In their natural condition," he wrote to his former student Carl Alvin Schenck in 1899,

> the Anglo-Saxon as a race hate systematic forestry. To some extent I have succeeded in educating Anglo-Indian opinion in that direction … what has to some extent educated them, has been the large net revenue, which we have been able to make.[42]

★★★

Regardless of Brandis's own skepticism, tropical forestry continued to evolve in line with British and then American possessions in Asia. In Malaya, Alfred Burn-Murdoch, a Scottish forester trained for a decade in Burma, was positioned as Chief Forest Officer in 1901, becoming Conservator of Forests in 1904 and organizing his department in accordance with Indian forestry administration lines. Farther east, the United States was testing its newly discovered expansionist ambitions in the attempt to organize the forests of the Philippines. The Insular Bureau of Forestry was established in 1900 under George Ahern, and forestry laws were passed in 1904 and 1905. Ahern was an army officer and avid explorer turned conservation enthusiast, whose affiliations with Gifford Pinchot got him the position and gave Pinchot access to the forest administration of the newly acquired colony.[43] Pinchot visited the Philippines in 1902, which provided him his first opportunity to experience tropical forests and closely examine a variety of trees, none of which he had known before.

But not too closely. As Greg Bankoff puts it, Pinchot's six-week journey through the islands was "forestry from the deck of the ship".[44] Some excursions were taken into the hinterland, but his observations on almost any subject were mostly speculative, clouded by his limited familiarity with the flora, soil composition, and climate of the archipelago. While Pinchot's input was instrumental in devising the rationale of the new administrative apparatus, and his optimism regarding the work in the Philippines placed it high in the achievements of what the "Anglo-Saxon people are trying to do", the actual organization of forests was sluggish for decades.[45] The difficulties in attracting professional foresters and adequate funding in its early years, the limited scientific knowledge Ahern possessed, and failures in understanding and incorporating local social and political conditions contributed to a very limited realization of its initial ambitions. The number of working plans grew slowly, and even as they were incorporated in the bureau's reports from 1906 onward, they were schematic and mostly non-applicable.[46] As a result, while extraction soared, forest cover sharply declined during the American period. In retrospect, several historical accounts regard the American forestry program in the Philippines as failure.[47]

As the American experience in the Philippines spiraled away from its initial intentions, the jungle regained vigor as a negative metaphor in tropical contexts. This was, as David Arnold suggests, related to an accumulation of scientific and colonial descriptions that consolidated various contexts under the monolithic heading of a tropical world circa 1950. Discussing French geographer Pierre Gourau's influential work, *Les pays tropicaux*, Arnold shows how this world was assigned with the old Buffonian inferiority of productive potential topped with later notions of pestilent and culturally backward environments that remain inherently hostile to the "modernization" efforts of northern civilizations.[48] Jungles are positioned in the frontlines of this struggle, and their subjugation remains a crucial strategic goal of Western world powers. Nowhere was this more evident than in the context of the grand environmental design project that took place during the American involvement in the Vietnam War.

The forests of Southeast Asia, hitherto unknown to most, became a quintessential American experience in the mid-1960s. In some regards, the war in Vietnam was not only a conflict between political ideologies but also a clash between different landscape mentalities. An individual growing up where the corn meets the sky could hardly be expected to navigate his way amongst the thickets of a tropical forest. Finding oneself in such an environment stood in sharp contrast to the experience of hiking through a forest in the United States, where the sublime may be found around the corner and a park ranger is always ready to lend a hand. The Vietnamese sylva was an alien landscape, missing all the cues and challenging reason. Stephen Wright, author of *Meditations in Green*, highlighted the rift when describing the GIs as "blind explorers", helplessly looking for their way in the shapeless jungle.[49] As platoons were optimistically marched into the green, supplied with the promise of superior tactics and the soundtrack of air cavalry, they encountered a hellish scenery of hindered navigation and guerilla warfare that cracked the rationale of American military organization. Individual and collective blindness were simultaneously a tactical fact and an existential state flirting with myth. When General William Westmoreland, commander of United States forces in Vietnam, likened the army to a "giant without eyes", he inadvertently revoked Polyphemus in his cave.[50] The Viet Cong were never compared to Odysseus and his entourage, but in using such metaphors, the war in Vietnam was recast as an epic struggle of the order of the Trojan War.[51] Epic challenges required epic military response, and Westmoreland's excessive firepower doctrine was up to the task. Calculating that the insurgents had no chance of winning a war of attrition, he made the mistake of looking at his data rather than at what was happening on the ground.[52] This was not exceptional under the tenure of Robert McNamara as Secretary of Defense, which was characterized by pious adherence to systems analysis and statistical modeling.[53] The crafting of Operation Rolling Thunder and other opulent displays of aerial superiority was fittingly focused on numbers – of planes, bombs, or dead bodies. While the metrics proved efficient when presented on mass media channels on the home front, these strategies did little to defeat the communist cause in South Vietnam.

But there was another line of strategic thinking that emerged in the early days of the conflict, based on close familiarity with trees and forests. The idea was to attack the jungle itself and so deny the insurgents forest cover. Its proponents hoped that if the jungle could be treated, the American army would get its sight back. Defoliation, as the treatment was termed, was a program that involved the latest developments in plant science, forestry work, and regional planning. At the same time, its turn to radical measures was mired in metaphorical discourse which perceived the forest as infested with hostile pathogens and vermin-like people. Following the pervasive metaphor of "the cosmic Southeast Asian jungle", Don Ringnalda writes that "in just about every Vietnam War novel one picks up, the author underscores America's adversarial relationship with the jungle."[54] He then makes the connection between the American distrust in jungles and defoliation, which "became a primary tactic in the war – almost a rite". By waging war upon the green, this fraught relationship sought resolution.

The outlines of jungle were not simply sketched by East–West dichotomies. In Vietnam, as in other parts of mainland Southeast Asia, the forested hills were often regarded as the refuge of dissenters and antagonizers, portrayed in negative terms in stark contrast to sedentary communities downstream. James C. Scott suggested that the region should be understood through an elevational cleavage: on the plains and deltas, states established ways of controlling settlements and increasing their tax base while on hills and mountains they encountered resistance, aversion, or indifference. The struggle was not only material but cultural: as valley states pushed for fixed boundaries and clear ethnic identities, highland communities displayed a jarring profusion of ethnicities, languages, and subsistence.[55] This complex human geography proved equally difficult for Vietnamese courts as it did for British colonial administrators in Burma. A common response in both cases would be attempts to civilize the "savages" on the hills.[56] The ruling elites in the Ngô Đình Diệm regime simply built on that history when perceiving the jungle as providing cover to its enemies and, by extension, hindering its modernization efforts. In this, American and Vietnamese readings of reality coalesced. William Colby, who served as the CIA's chief of station in Saigon and was close to Diệm and his circle, described the 1960 creation of the National Liberation Front (NLF) as an ominous event taking place in a jungle.[57] It is not surprising then that the first systematic attempts to alter the Vietnamese jungle environment were collaborative efforts by the Diệm and US administrations. In late 1961, President Kennedy signed a memorandum authorizing defoliation missions in South Vietnam, setting off a decade of unprecedented landscape alterations and environmental havoc. However, at this initial phase, spraying of the jungle was regarded as a "selective and carefully controlled joint program", working closely with the South Vietnamese in identifying key routes to be cleared. Spraying crops would be considered only if "the most careful basis for resettlement and alternative food supply has been created."[58] In other words, defoliation was not only initiated in collaboration with the Diệm regime but was in tandem with its ambitious replanning of the agrarian hinterland.

By 1959, the disruptive influence of anti-government forces was very present in South Vietnam. As a response, Diệm sought to counter the influence of communist ideas on the peasantry by introducing several far-reaching resettlement programs that shifted gears from earlier attempts at land reform.[59] The initial idea was to relocate groups that were particularly susceptible to North Vietnamese influence in "agglomeration centers" of one type (*qui khu*) and concentrating groups sympathetic with the government in another type (*qui ap*). By this spatial reorganization along political lines, it was hoped that the government would be able to better control the insurgents and protect its supporters. For several months, Saigon exerted considerable pressure on local officials to adopt the program but was immediately confronted with bitter resistance. The quick disintegration of this crude planning attempt demonstrated the inherent difficulty of telling friends from enemies in the complex social realities of South Vietnam, one that will plague all subsequent plans. The Diệm government soon began to formulate a new, more comprehensive plan which was presented in July. This plan was based on *agrovilles*, which the president described in a speech as "densely populated settlement areas in the countryside, where conditions are favorable to communication and sanitation". Quickly adapting the rhetoric in response to the failure of the previous scheme, Diệm posed that the agrovilles "will not only improve the life of the rural population, but they will also constitute the economic units which will play an important role in the future development of the country as a whole."[60]

The agroville plan focused on the Mekong Delta, proposing twenty-four model settlements of 2,000–3,000 inhabitants, each concentrating an area's rural population on small plots while allowing them to retain their existing land outside the settlement. The number of agrovilles was expected to grow, affecting 500,000 peasants. Diệm modeled these new towns on earlier schemes of "New Villages" in Malaya, which relocated a similar number of people, and at the same time insisted that beyond security considerations, the population would also benefit from a higher standard of living. In reality, the building of the first agroville of Vị Thanh – Hỏa Lựu, which Diệm inaugurated in April 1960, demonstrated many of the difficulties ahead. Conceived as the largest settlement of this type and model for all others, the settlement was partly designed by Ngô Viết Thụ, a well-known Vietnamese architect who became close with Diệm and would later design Independence Palace in Saigon. A dispatch from the American ambassador described the plan as projecting "four separate but contiguous villages – three of 200 hectares each south of the road-canal, and one of 400 hectares north of the road-canal and opposite the central village". He went on to describe the buildings as "substantial as well as pleasing to the eye", complemented by "elaborate artificial lakes", which "add beauty to the city centers".[61] However, what sounds like a description to a new American suburb was based on free labor, in which 20,000 peasants were forced by local officials to dig the canals and build the city in no time during harvest season. The palace in Saigon was largely blind to these practices, convinced as it was that the plan's benefits were clear to all those who were supposed to build it.[62] The ambitious plan quickly lost traction, and by the end of 1960, Diệm informed the Americans that it would be halted. Soon after, the palace moved to the security-oriented Strategic Hamlet Program, which attempted to relocate populations in conflict zones into small defensive settlements, almost medieval in rationale and appearance. Phillip Catton noted that "the progress or otherwise of the Saigon regime can be charted by a series of vast rural schemes, which would culminate in the Strategic Hamlet program".[63] However, even in their most elaborate civic moments, the thrust of these schemes was fueled and largely defined by counterinsurgency and, by extension, the association made by the Diệm regime between unruly jungles and anti-nationalist guerilla forces.[64] A 1961 presentation of the Ban The' agroville, located about ninety kilometers northeast of Vị Thanh, is emblematic in this regard: a first scale model, created as part of the propaganda campaign surrounding the new settlement, shows the "before" condition as an undefined swath of forest covering the low Ba The' mountain and its environs. The "after" model presents a neatly contained forest topography surrounded on all sides by roads, along which rows of small parcels stretch, complete with houses and gardens. Public buildings dot the scheme here and there. Another wooded area, possibly meant for communal use, is contained within a circular road. The realized scheme was even more extreme, serving as a perfect illustration of Scott's argument about the conclusion of the elevation-based conflict between state apparatuses and hill communities in the late twentieth century. It features concentric ring roads and canals encircling the mountain, both physically isolating it from the larger forest and symbolically turning it into a vestige of insurgent potential (Figures 2.6 and 2.7).

The early spraying missions that followed Kennedy's memorandum were regarded by both the Americans and the Vietnamese as complementary to the same rationale. By focusing on main arteries that connected existing and projected settlements, the flights could easily be counted as part of the ambition to clear the way as means to develop a modern infrastructure for the South Vietnamese hinterland and integrate it fully into a functioning state system. However, a confluence of momentous events soon diverted its trajectory. The history of the herbicidal war in Vietnam is one of the most notorious episodes in modern warfare and was described in great detail based on ample primary sources that exist on the

FIGURE 2.6   An "after" model of Ba The' Agroville, Vietnam, 1962. Harry S. Truman Library.

American and, to a lesser extent, the Vietnamese sides. These descriptions include meticulous chronologies of the figures and agencies involved, their contextualization within larger movements in military and scientific thinking, and the role Operation Ranch Hand played in the mobilization of international alliances and public opinion.[65] Considered as a megaproject of forest design, its outline can be sketched along intertwining lines of scientific experimentation and military thinking.

The science of defoliation developed during the early decades of the twentieth century through research into synthetic growth hormones. Its military potential was recognized during World War II by plant scientists in American and British contexts that envisioned its application on forests in the Pacific front. Ezra Kraus, chair of the University of Chicago's Department of Botany who pioneered this line of thinking, embroiled forests with antagonistic qualities when writing that the "distribution of sprays or mists over enemy forests would, through killing of trees, reveal concealed military depots".[66] This assumption, reached at in the botany lab before any real research was conducted on "enemy" forests, appealed to the military, which funded further investigations. As research into weed-killing phenoxyacetic compounds progressed, tests began in the mid-1940s with retrofitted airplanes flying on spraying missions over the Florida Everglades, chosen for their assumed resemblance to tropical

FIGURE 2.7   Aerial view of Ba The' Agroville, Vietnam, 1962. Harry S. Truman Library.

forests in the front. The chemicals did not make it in time before the war was decided by nuclear means, but they spurred an arms race between chemical companies looking to feast on the booming demand for weed killers in forestry, agriculture, and postwar housing developments. Much like its contemporary DDT, the sales of 2,4-D, the most popular phenoxy herbicide of the time, increased by hundreds of percent within a few years, propelled by well-circulated propaganda about weeds that cause billions of dollars in annual losses across the country. The introduction of hundreds of market products and the high investment in research and development allowed the chemical companies a smooth transition into the war business once the need arose on the Southeast Asian front. Mixing 2,4-D with the more aggressive 2,4,5-T and cooking it in concentrations up to twenty times higher than those used on American fields and forests, the companies created Agent Orange as a super defoliant that could "improve vertical and lateral vision in forested terrain"[67] and expose NLF insurgents to the elements of carpet-bombing.

According to American sources, President Diệm was highly support-ive of the defoliation program, which the Pentagon labored to present as a joint venture. To make it clear, the Americans emphasized Diệm's involvement in choosing the targets in early spray missions.[68] This sup-posedly harmonious beginning was soon challenged by more extreme propositions raised within the administration by officials who were quick to conclude from the limited evidence at hand that it was possible to expand spraying reach to the entire geographical and ecological spec-trum of Vietnamese forests.[69] The growing military delusion informed decisions to attack more and more areas, still in coordination with the Vietnamese army. However, as both Diệm and Kennedy were assassinated in late 1963, the link between defoliation and pacification was severed. As the war was "Americanized" under Lyndon Johnson and American troops began to flow in in large numbers, defoliation missions increased in frequency and ambition. By the curtailing of Operation Ranch Hand in 1971, approximately 12 percent of South Vietnam would be sprayed, making it one of the most dramatic attempts at deliberate landscape altera-tions in history. Some environmental historians, taking a sweeping glance at the effects of military campaigns on natural systems, had compared the havoc caused in the coastal mangroves and the highland rainforests in Vietnam to the destruction of southern landscapes in the American Civil War or of European ones in the Western Front during World War I.[70] The fundamental difference between Operation Ranch Hand and these earlier examples is the fact that in Vietnam, the aggression was directed specifi-cally at foliage, and the damage to humans – both Vietnamese civilians and American soldiers – was conceived as collateral. It was a botanical war more akin to forestry operations than to conventional combat.

Yet it was conducted for a long time with only a scant presence of forest scientists. Among the experts in early herbicide use in US labs and on the ground in South Vietnam, none specialized in or focused primarily on forests. The lack of such knowledge and the plant scientists' limited capacity to scale up the effect of chemicals from individual trees to the tropical forest system was evident in some of the spraying program's self-defeating outcomes. Even the most basic assumption of Ranch Hand, that the spraying would improve vertical and lateral vision, was difficult to ascertain once projected onto the forests. Ironically, defoliation activities changed the sectional balance of power within the forest: as herbicides infiltrated and mutilated the higher trees through their canopies, more sunlight reached the forest floor, promoting the growth of plants that took over the terrain. This, in turn, made defoliated areas more efficient as camouflage for NLF troops and more difficult for US ground forces. That is not to suggest that the scientists at the helm were not attuned to the nuances of the Vietnamese context. Reports that were sent back soon after the first spraying missions in early 1962 already highlighted the different nature of the mangroves along the southern coast and the evergreen forests of the central highlands and Mekong Delta. The authors also noted that the effectiveness of spraying missions would vary according to season and time of day.[71] But these hard-won observations were in many ways repeating the ones made in nineteenth-century India and then in Malaya and the Philippines, where imperial foresters were ignoring gaps in scientific knowledge when trying to establish management patterns that could distill economically useful trees from vast areas of forest undergrowth. Forest management thus became a major concern, which turned Ranch Hand into a self-perpetuating mission: since the lasting effectiveness of defoliation was subject to the cycles of both plant growth and guerilla movement, US military commanders reasoned more areas should be sprayed more frequently with more chemicals. At the height of the operation, as vast quantities of chemicals were dropped on forests and fields, concrete scientific input played an even smaller role in the choice of targets. Whether in the large-scale spray missions and the smaller sorties meant to address a tactical problem, it was forest design from the deck of the plane.[72] Consequently, the "working plans" were, at best, rudimentary.

The total war declared on the jungle continued unabated until about 1970, when science reentered the scene and began to document and communicate the disastrous impacts of "preventing a forest".[73] At this time, forest perspectives were instrumental in bringing to light the long-term environmental and health legacies that would plague inflicted regions for decades to come. As the scientists were piling evidence on the ground in preparation for their later assault on American decision-making processes, some of the more critical voices among them sought to bring the magnitude of environmental destruction to wider public attention. Defoliation was already featured before in the rhetoric of environmentalist groups and individuals, most notably during the inaugural Earth Day in April 1970.[74] Yet its importance was compounded later that year as documentation, which included aerial photographs gathered from the army and taken by the scientists, was quickly diffused in popular science journals and books.[75] The aerial photos were particularly striking for several reasons. First was the exposing of the overwhelming scale of herbicidal operations, which took out entire areas and transformed them into zombie forests. Photographs of sprayed and unsprayed coastal mangroves, taken from similar angles and published side by side, delivered the point by giving the impression of before and after documentation. The scale not only was a matter of sprayed area but was also related to the indefinite time it would take before natural systems could recover, if at all. Second, the photographs made clear the designed nature of the military endeavor. While it may have been hard to gather what was happening on the ground based on the partial news reports and the strict classification of Operation Ranch Hand available to the public, images of forest areas ending in clear-cut lines and perforated by bomb craters made it clear that this was not a fog-of-war situation but a preconceived program of landscape transformation. Third, the images of wrecked forests in South Vietnam strangely rhymed with memories of denuded American landscapes, recently destroyed by overlogging and overcapitalizing, and still present in the American collective consciousness. Thus, the jungle acquired an uncanny familiarity and was seamlessly integrated into the rhetoric against capitalist overexploitation at home and abroad. With that, a tectonic shift in the long-standing metaphor of the jungle began to take place. The dark sylvan territory of unfathomable danger, the disease-stricken green hell that hinders human bodies and civilizations was increasingly recast through its potential for human flourishing, even salvation. While not always proceeding conclusively, the early years of the 1970s mark the birth of the current conception of tropical forests as spaces of biological diversity, environmental potential, ancestral wisdom, and more sustainable and moral ways of inhabiting the earth (Figure 2.8).

Take Ursula Le Guin's novella *The Word for World Is Forest*, written at the height of anti-war demonstrations and with defoliation missions in mind.[76] The story might have taken place in Indochina: it is a world entirely made of forest, in which human colonizers from planet Terra find

FIGURE 2.8   Aerial photographs of a portion of mangrove forest and deforested mangrove forests in Vietnam, 1970, taken by Matthew Meselson. Courtesy of the Meselson Archive.

out that the inhabitants – the short and slow humanoids they affection-lessly call "creetchies" – are capable of inflicting lethal damage through intelligent organization and guerilla-style resistance. Here, too, contrast-ing landscape views drive the narrative, but the hostility toward forest that is pronounced by Captain Davidson, the Terran soldier anti-hero, is answered by its descriptions by Selver, the indigenous protagonist who reckons the forest's inherent complexity:

> No way was clear, no light unbroken, in the forest.... The view was never long, unless looking up through the branches you caught sight of the stars. Nothing was pure, dry, arid, plain. Revelation was lacking. There was no seeing everything at once: no certainty.[77]

The novella meticulously describes the planet's native societies, dwelling on their physical and mental entanglement with their environment:

> the voices calling here and there and the babble of women bathing or children playing down by the stream, were not so loud as the morning birdsong and insect-drone and under-noise of the living forest of which the town was one element.[78]

Such descriptions stroke a deep chord with generations of readers and academics who saw them through their own changing concepts of society, culture, and politics.[79]

For centuries, the metaphor of the unruly forest carved meandering paths through time and space, unearthing strange connections between Gauls and Hmong, between naturalists and colonial administrators, between medieval kingdoms and communist states, between North American suburbs and Southeast Asian new towns. However, by the 1970s, it was mostly bygone. The most remote points on the planet were reached and mapped by new infrastructures and new satellites. Combined with a new consciousness regarding the planet's limited resources and unlimited complexity, these transformations paved a route through the heart of darkness, not only making remote forests familiar but implicitly suggesting them also as models for living. The benevolent jungle thus found its way into the schemes of planners and architects seeking to intro-duce ecological ideas into their work. The forest premonitions of Mies van der Rohe were perhaps signaling something different than the ulti-mate defeat of humankind.

Notes

1 Quoted from an interview with John Peter, in Mertins, Detlef. "Living in a Jungle: Mies, Organic Architecture, and the Art of City Building." In *Mies in America*, edited by Phyllis Lambert. Montréal: Canadian Centre for Architecture; New York: Whitney Museum of American Art; Harry N. Abrams, 2001, 618.

2 Mertins, Detlef. "Living in a Jungle: Mies, Organic Architecture, and the Art of City Building." In *Mies in America*, edited by Phyllis Lambert. Montréal: Canadian Centre for Architecture; New York: Whitney Museum of American Art; Harry N. Abrams, 2001, 618.

3 Mertins, Detlef. "Living in a Jungle: Mies, Organic Architecture, and the Art of City Building." In *Mies in America*, edited by Phyllis Lambert. Montréal: Canadian Centre for Architecture; New York: Whitney Museum of American Art; Harry N. Abrams, 2001, 618.

4 Quoted in Boyarsky, Nicholas. "Review of In the Shadow of Mies". *AA Files*, no. 22 (1991): 106–7.

5 Caldwell was a landscape architect trained at the office of Jens Jensen, who developed a distinct organic approach to landscape architecture. He joined the IIT faculty in 1944 and allegedly ghostwrote some of Mies's speeches while rendering Hilberseimer's plans. For a colorful account of Caldwell's years at IIT, see Dyja, Thomas. *The Third Coast: When Chicago Built the American Dream*. New York: The Penguin Press, 2013.

6 On the sources of these fantasies, see Zantop, Susanne. *Colonial Fantasies: Conquest, Family, and Nation in Precolonial Germany, 1770–1870*. Durham, N.C.: Duke University Press, 1997.

7 Stanley's book is filled with fantastic and chilling literary descriptions of jungle suffering. For instance, when entering the "black and chill forest called Mitamba … [we were] bidding farewell to sunshine and brightness". Inside, the world changes:

> all this time the trees kept shedding their dew upon us like rain in great round drops. Every leaf seemed weeping. Down the boles and branches, creepers and vegetable cords, the moisture trickled and fell on us. Overhead the wide-spreading branches, in many interlaced strata, each branch heavy with broad thick leaves, absolutely shut out the daylight. We knew not whether it was a sunshiny day or a dull, foggy, gloomy day; for we marched in a feeble solemn twilight, such as you may experience in temperate climates an hour after sunset. The path soon became a stiff clayey paste, and at every step we splashed water over the legs of those in front, and on either side of us.

These descriptions often mix Humboldtian impressions with atmospheric fear:

> To our right and left, to the height of about twenty feet, towered the undergrowth, the lower world of vegetation. The soil on which this thrives is a dark-brown vegetable humus, the debris of ages of rotting leaves and fallen branches, a very forcing-bed of vegetable life, which, constantly fed with moisture, illustrates in an astonishing degree the prolific power of the warm moist shades of the tropics.
>
> *In Stanley, Henry M.* Through the Dark Continent; Or, The Sources of the Nile, Around the Great Lakes of Equatorial Africa, and down the Livingstone River to the Atlantic Ocean. *London: Sampson Low, Marston, Searle & Rivington, 1879, 408–10*

8 Hemming, John. *Amazon Frontier: The Defeat of the Brazilian Indians.* Cambridge, Mass.: Harvard University Press, 1987, 471.

9 Penny, H. Glenn. "The Politics of Anthropology in the Age of Empire: German Colonists, Brazilian Indians, and the Case of Alberto Vojtěch Frič." *Comparative Studies in Society and History* 45, no. 2 (2003): 249–80.

10 Conrad, Sebastian. *Globalisation and the Nation in Imperial Germany.* Cambridge; New York: Cambridge University Press, 2010, 294.

11 Pratt, Mary Louise. *Imperial Eyes: Travel Writing and Transculturation.* London; New York: Routledge, 1992, 18. For a thorough account of the expedition, recounting some of the literature style involved, see Crane, Nicholas. *Latitude: The True Story of the World's First Scientific Expedition.* New York: Pegasus Books, 2021.

12 Condamine, Charles-Marie de la. *A Succinct Abridgment of a Voyage Made Within the Inland Parts of South-America: From the Coasts of the South-Sea, to the Coasts of Brazil and Guiana, Down the River of Amazons: As It Was Read in the Public Assembly of the Academy of Sciences at Paris, April 27, 1745.* London: E. Withers, 1747, iv.

13 Buffon claimed that the differences between groups are derived from environmental conditions, implying that racial characteristics may be altered over time. On Buffon's ideas about race, and how they contributed to the scientific racial discourse that followed, see Hudson, Nicholas. "From Nation to 'Race': The Origin of Racial Classification in Eighteenth-Century Thought." *Eighteenth-Century Studies* 29, no. 3 (1996): 247–64.

14 Buffon, Georges Louis Leclerc. *Buffon's Natural History: Containing a Theory of the Earth, a General History of Man, of the Brute Creation, and of Vegetables, Minerals, &c. &c.* London: Printed for the Proprietor, and sold by H. D. Symonds, 1807, vol. VII, 45.

15 Gerbi, Antonello, and Jeremy Moyle. *The Dispute of the New World: The History of a Polemic, 1750–1900.* Translated by Jeremy Moyle. Pittsburgh, Pennsylvania: University of Pittsburgh Press, 1973, 40.

16 Some of the writers influenced by Buffon took it upon themselves to portray an abysmal image of America, most notably in Cornelius de Pauw's controversial and influential book *Recherches philosophiques sur les Américains*, which included outrageous statements about the sterileness of the continent, its stupefying effects on natives, the natives' inferiority in culture and intellectual faculties, and the detrimental environmental effects that turned Europeans into cannibals.

17 Brandis, Dietrich. *Indian Forestry*. Woking: Oriental University Institute, 1897, 49.

18 Schenck, Carl Alwin. *The Biltmore Story*. Santa Cruz, Calif: Forest History Society, 1955, 9.

19 Marshman was the daughter of Joshua Marshman, a prominent Baptist missionary in Bengal who translated the Bible into Indian languages, as well as the Ramayana into English, and founded the Serampore College. She was an avid botanist who wrote the introduction to her first husband's posthumously published extensive catalogue of the East India Company's botanical gardens in Calcutta and Serampore. See Voigt, J. O. *Hortus Suburbanus Calcuttensis: A Catalogue of the Plants Which Have Been Cultivated in the Hon. East India Company's Botanical Garden, Calcutta, and in the Serampore Botanical Garden, Generally Known as Dr. Carey's Garden, from the Beginning of Both Establishments (1786 and 1800) to the End of August 1841*. Dehra Dun, India: Ravendra Pal Singh Gahlot, 1984.

20 Hesmer, Herbert. *Leben Und Werk von Dietrich Brandis 1824–1907*. Wiesbaden: Springer Fachmedien Wiesbaden, 1975, 10–12.

21 The dramatic shift in power followed the Indian Rebellion of 1857, after which doubts about the company's ability to properly manage the Indian territories surged in the English political and public spheres.

22 From an 1850 letter to the Agri-Horticultural Society of Lahore. Quoted in Grove, Richard. *Green Imperialism: Colonial Expansion, Tropical Island Edens, and the Origins of Environmentalism, 1600–1860*. Cambridge; New York: Cambridge University Press, 1995, 455.

23 Grove, Richard. *Green Imperialism: Colonial Expansion, Tropical Island Edens, and the Origins of Environmentalism, 1600–1860*. Cambridge; New York: Cambridge University Press, 1995, 455.

24 The Rashomon nature of the first survey to Pegu is further complicated by a third account, made by Robert Abreu, assistant to the superintendent of forests in Pegu, which describes almost none of the epic hardships, portraying instead a calm and almost romantic voyage through native settlements, picturesque ruins of past civilizations, and serene agricultural and forestlands. See Abgeu, Robert. *Journal of a Tour Through the Pegu & Martaban Provinces in the Suite of Drs. McClelland & Brandis*. Maulmain: T. Whittam, 1858.

25 Brandis, Dr. *Report on the Teak Forests of Pegu, with a Memorandum on the Teak in the Tharawaddy Forests*. London: Eyre and Spottiswoode, 1856, 9.

26 Brandis notes that available science is not sufficient even in ascertaining the age of teak trees. Brandis, Dr. *Report on the Teak Forests of Pegu, with a Memorandum on the Teak in the Tharawaddy Forests*. London: Eyre and Spottiswoode, 1856, 11.

27 Brandis, Dr. *Report on the Teak Forests of Pegu, with a Memorandum on the Teak in the Tharawaddy Forests*. London: Eyre and Spottiswoode, 1856, 16.

28 The report was written by William Campbell Onslow, a superintendent in Coorg Province, and described Coomree cultivation in Mysore, which Brandis found similar enough to include it in his own report. Brandis, Dr. *Report on the Teak Forests of Pegu, with a Memorandum on the Teak in the Tharawaddy Forests*. London: Eyre and Spottiswoode, 1856, 34.

29 Sivaramakrishnan, Kalyanakrishnan. "Colonialism and Forestry in India: Imagining the Past in Present Politics." *Comparative Studies in Society and History* 37, no. 1 (1995): 7.

30 Sivaramakrishnan, Kalyanakrishnan. "Colonialism and Forestry in India: Imagining the Past in Present Politics." *Comparative Studies in Society and History* 37, no. 1 (1995): 14. For a detailed history of local resistance to these measures, see Guha, Ramachandra. *The Unquiet Woods: Ecological Change and Peasant Resistance in the Himalaya*. Berkeley: University of California Press, 1990.

31 On the creation of forest administration and the involvement of German forest scientists in Java, see Peluso, Nancy Lee. "The History of State Forest Management in Colonial Java." *Forest & Conservation History* 35, no. 2 (1991): 65–75. On German forest management in Tanganyika, see Schabel, Hans G. "Tanganyika Forestry under German Colonial Administration, 1891–1919." *Forest & Conservation History* 34, no. 3 (1990): 130–41. On the strict regulation of farming in French Indochina, see Ngô, Vinh Long. *Before the Revolution: The Vietnamese Peasants under the French*. Cambridge, Mass.: MIT Press, 1973.

32 "It is recorded that actually the first plantation to be so made was formed by U Panhee, a Karen in the Thonze forests, as a personal present to Brandis." Blanford, H. R. "Highlights of one Hundred Years of Forestry in Burma." *Empire Forestry Review* 37, no. 1 (91) (1958): 34.

33 On this subject, see Munshi Saldanha, Indra. "Colonialism and Professionalism: A German Forester in India." *Economic and Political Weekly* 31, no. 21 (1996): 1265–73.

34 Guha, Ramachandra. "The Prehistory of Community Forestry in India." *Environmental History* 6, no. 2 (2001): 223.

35 The working plan became a standard document in India's forest management, especially after Wilhelm Schlich succeeded Brandis as Inspector General in 1873. Over the course of his tenure, plans gradually became more sophisticated in their ability to implement feedback from the ground and change over time. See Barton, Greg. *Empire Forestry and the Origins of Environmentalism*. Cambridge; New York: Cambridge University Press, 2002, 81.

36 Latter, identified by initials on the map, was an army official who participated in the seizure of Pegu. He acted as superintendent of forests and submitted several botanical reports on the area. Additionally, he held the position of deputy commissioner in Prome, where he was killed in his sleep in 1853.

37 Engaben was a commonly used name for a species of the genus *Dipterocarpus*, yielding kreuing timber.

38 Brandis notes that when he first visited the provinces of British Burma in the mid-1850s, more than 90 percent was forested. Brandis, *Indian Forestry*, 30.

39 One should not assume that the Karens were given rights in the same way indigenous stewardship is being discussed today. Their cultivation was to be supervised by forest officials, who represented and promoted colonial interests.

40 There is a good reason to believe Brandis influenced the forest work at Biltmore. Brandis's work in India was mentioned in *Garden and Forest*, which Olmsted must have known, in March 1888. A few years later, he would publish a series of articles on the forests of Burma in that same publication. See *Garden and Forest*, v. 1, March 21, 1888. Gifford Pinchot studied with Brandis in France in 1889 and upon returning was advised by Brandis to find a large private landowner who would be interested in advancing the cause of scientific forestry. Pinchot found that owner in George Washington Vanderbilt and got himself hired by Olmsted in 1891. See Roper, Laura Wood. *FLO: A Biography of Frederick Law Olmsted*. Baltimore: Johns Hopkins University Press, 1973, 418.

41 This narrative was bulwarked by Pinchot, who wrote:

> Here was my chance, Biltmore could be made to prove what America did not yet understand, that trees could be cut and the forest preserved at one and the same time … thus Biltmore Forest became the beginning of practical forestry in America. See Pinchot, Gifford. *Breaking New Ground*. Seattle: University of Washington Press, 1972 [1947], 49.

On "Empire Forestry", see Barton, *Empire Forestry and the Origins of Environmentalism*.

42 Quoted in Munshi Saldanha, "Colonialism and Professionalism", 1271.

43 Rakestraw, Lawrence. "George Patrick Ahern and the Philippine Bureau of Forestry, 1900–1914." *The Pacific Northwest Quarterly* 58, no. 3 (1967): 142.

44 Bankoff, Greg. "Breaking New Ground? Gifford Pinchot and the Birth of 'Empire Forestry' in the Philippines, 1900–1905." *Environment and History* 15, no. 3 (2009): 371.

45 Pinchot quoted by Bankoff. Bankoff, Greg. "Breaking New Ground? Gifford Pinchot and the Birth of 'Empire Forestry' in the Philippines, 1900–1905." *Environment and History* 15, no. 3 (2009): 371.

46 The earliest of these plans, for a "public forest tract" of the Insular Lumber Company, was more a marketing tout than a workable plan. See Everett, H. D., and Harry Nichols Whitford. "A Preliminary Working Plan for the Public Forest Tract of the Insular Lumber Company: Negros Occidental." In *Report of the Philippine Commission. Pt. 3*, Washington: G.P.O., 1906, 651–78.

47 Beyond Bankoff's highly critical account, see also Luyt, Brendan. "Empire Forestry and Its Failure in the Philippines: 1901–1941." *Journal of Southeast Asian Studies* 47, no. 1 (2016): 66–87, and Roberts, Nathan E. "U.S. Forestry in the Philippines: Environment, Nationhood, and Empire, 1900–1937." PhD diss., University of Washington, 2014.

48 The book was originally published in 1947 and appeared in English as "The Tropical World" in 1953, with two revised editions published during the following decade. See Arnold, David. "'Illusory Riches': Representations of the Tropical World, 1840–1950". *Singapore Journal of Tropical Geography* 21, no. 1 (March 2000): 6.

49 Wright, Stephen. *Meditations in Green.* New York: C. Scribner's Sons, 1983, 76.

50 Westmoreland, William C. "Westmoreland on the Army of the Future." *NACLA Newsletter* 3, no. 7 (1969): 14–16.

51 Carl von Clausewitz made a similar argument in *On War,* warning armies that forests may cast them in the role of fighting "like a blind man against one with his eyesight". See Clausewitz, Carl von. *On War.* New Jersey: Princeton University Press, 1989 [1832], 452.

52 The change of strategy came about in mid-1965, as the American leadership supposed that the insurgency had little support from the South Vietnamese population and that attrition could be used to push the northern regime out of the war, drying up the southern resistance. See Clarke, Jeffrey. "On Strategy and the Vietnam War," *Parameters: Journal of the US Army War College* 16, no. 1 (1986): 40.

53 Journalist David Halberstam famously pointed to this chronic disparity between facts and numbers in the early 1970s. Later it would be termed "the McNamara fallacy". See Halberstam, David. *The Best and the Brightest.* New York: Random House, 1972.

54 Ringnalda, Don. *Fighting and Writing the Vietnam War.* Jackson: University Press of Mississippi, 1994, 51.

55 Scott proposes that this cleavage was unfolding for millennia. Scott, James. *The Art of Not Being Governed: An Anarchist History of Upland Southeast Asia.* New Haven: Yale University Press, 2009, 5.

56 Patricia Pelley writes that "civilization could be gauged by geography and, more specifically, by elevation. The people in the lowlands (i.e., ethnic Vietnamese) were fully civilized … but highlanders were still savage, and the higher the elevation the greater the degree of savagery." Pelley, Patricia M. *Postcolonial Vietnam: New Histories of the National Past.* Durham: Duke University Press, 2002, 89.

57 "In its best conspiratorial style, a clandestine radio station announced on December 12 the formation in a South Vietnamese jungle of a 'National Front for the Liberation of South Vietnam,' which was finally confirmed in a Hanoi broadcast of January 29, 1961." In Colby, William, and James. McCargar. *Lost Victory: A Firsthand Account of America's Sixteen-Year Involvement in Vietnam.* Chicago: Contemporary Books, 1989, 81.

58 Bundy, McGeorge. "National Security Action Memorandum No. 115: Defoliant Operations in Viet Nam," November 30, 1961, to the Secretary of State, The Secretary of Defense, JFKNSF-332-017, John F. Kennedy Presidential Library and Museum.

59 Diệm had argued for the need for land redistribution in the Mekong Delta, which suffered from an overwhelming concentration of ownership, an inheritance from the French colonial rule. Yet the legislation was extremely conservative and eventually did not achieve its goals. His administration later pursued large scale countryside development schemes meant, among other things, to assimilate ethnic minorities into the Vietnamese nation. See Ngô, *Before the Revolution*, 43–60; Woodside, Alexander B. *Community and Revolution in Modern Vietnam*. Boston: Houghton Mifflin, 1976, 120–24; and Hickey, Gerald Cannon. *Free in the Forest: Ethnohistory of the Vietnamese Central Highlands, 1954–1976*. New Haven: Yale University Press, 1982, 5–12.

60 Quoted in Zasloff, Joseph J. "Rural Resettlement in South Viet Nam: The Agroville Program." *Pacific Affairs* 35, no. 4 (1962), 327.

61 Durbrow, Eldrige. "GVN Agroville Program." In *Foreign Relations of the United States, 1958–1960, Vietnam, Volume I*, edited by Edward C. Keefer and David W. Mabon, Document 169, Despatch from the Ambassador in Vietnam to the Department of State, No. 426, Saigon, June 6, 1960. Washington: Government Printing Office, 1986.

62 The dependence on local labor was part of the Diệm administration's wish to realize the program without US aid and was also linked to its ideological mission of "community development", which posed that unpaid communal labor would build collective character. See Zasloff, *Rural Settlement*, 334.

63 Catton, Philip E. *Diệm's Final Failure*. Lawrence: University Press of Kansas, 2002, 219.

64 The contrast between hill forests and the modern state was shared by the DRV in North Vietnam, which initiated during the same time mobilization campaigns to "sedenterize the nomads" and "storm the hills", in which ethnic Vietnamese villagers were resettled in higher altitudes in order to transform the region. Pelley, *Postcolonial Vietnam*, 99.

65 For the official chronology, based on military sources, see Buckingham, William A., and United States. Air Force. Office of Air Force History. *Operation Ranch Hand: The Air Force and Herbicides in Southeast Asia, 1961–1971*. Washington, D.C.: Office of Air Force History, United States Air Force, 1982. For a discussion about the role of scientists in military developments within the context of the Cold War, see Bridger, Sarah. *Scientists at War: The Ethics of Cold War Weapons Research*. Cambridge, Mass.: Harvard University Press, 2015. For a striking account of the impact of chemical war on environmental thinking within the sciences and as a matter of public policy, see Zierler, David. *The Invention of Ecocide: Agent Orange, Vietnam, and the Scientists Who Changed the Way We Think About the Environment*. Athens [Ga.]: University of Georgia Press, 2011. For the massive influence of herbicide use on public protests, see Hay, Amy Marie. *The Defoliation of America: Agent Orange Chemicals, Citizens, and Protests*. Tuscaloosa: The University of Alabama Press, 2022.

66 Quoted in Zierler, *The Invention of Ecocide*, 40.

67 Zierler, *The Invention of Ecocide*, 5. The higher concentrations and hasty production process became a heated subject in years to come, as its effects on plants, animals, and humans were never tested before, and as it introduced to sprayed territories vast quantities of dioxin, a toxic by-product of the cooking with a lasting environmental footprint.

68 Zierler, *The Invention of Ecocide*, 59.

69 Such propositions, designed to "support Diệm's administration", were circulating as early as 1961. See Zierler, *The Invention of Ecocide*, 61.

70 See, for instance, McNeill, John. "Woods and Warfare in World History." *Environmental History* 9, no. 3 (July 2004): 388.

71 Zierler, *The Invention of Ecocide*, 76.

72 Strategic large-scale defoliation had to go through both diplomatic and military chain of command before approval, and planned months in advance. Smaller missions were approved on a local level in thirty to ninety days. See 14th Military History Detachment. "Chemical Support within the 1st Cavalry Division – 25 Jan. 1971." 1971. Record Group 472: Records of the U.S. Forces in Southeast Asia 1950–1976. National Archives Identifier (NAID): 57588941. National Archives at College Park, Maryland.

73 The army's anti-jungle rhetoric aptly bootlegged a Smokey Bear ad for fire prevention and turned into the unofficial slogan of Ranch Hand: "only you can prevent a forest".

74 Denis Hayes, environmental advocate and coordinator of the first Earth Day, made the connection during his speech in Washington: "We have made Vietnam an ecological catastrophe … we dumped defoliants on Vietnam at a rate of 10,000 pounds a month… we cannot pretend to be concerned with the environment … as long as we continue the war." Hayes, Denis. "The Beginning." *Give Earth a Chance: Environmental Activism in Michigan.* Accessed January 8, 2024. https://michiganintheworld.history.lsa.umich.edu/environmentalism/items/show/759.

75 Matthew Meselson and Arthur Westing were probably the most effective scientists in popularizing the subject. Following their return from the American Association for the Advancement of Science (AAAS) Herbicide Assessment Commission in Vietnam in August 1970, they quickly published many articles and books exposing and assessing the effects of herbicidal war. These publications used provocative titles such as *The Cratering of Indo-China* (in *Scientific American*) and *Harvest of Death* (co-authored book), lending urgency and dramatic effect to the issue.

76 In an introduction to the 1976 book edition, LeGuin writes:

> It was becoming clear that the ethic which approved the defoliation of forests and grainlands and the murder of noncombatants in the name of 'peace' was only a corollary of the ethic which permits the despoliation of natural resources for private profit or the GNP, and the murder of the creatures of the Earth in the name of 'man.'. See Guin, Ursula K. Le. *The Word for World Is Forest.* New York: Berkley Books, 1976 [1972].

77 Guin, Ursula K. Le. *The Word for World Is Forest*. New York: Berkley Books, 1976 [1972], 26.

78 Guin, Ursula K. Le. *The Word for World Is Forest*. New York: Berkley Books, 1976 [1972], 40.

79 The Novella, written at a time when LeGuin was still considered a genre writer, was reprinted numerous times since and widely discussed in academic writing in relation to various themes and perspectives, from cultural theory to postcolonial studies and from anthropology to imperialist nostalgia in the film *Avatar*.

3

THE THOUSAND-YEAR FOREST

In the aftermath of the particularly cold winter of 2021, there appeared a specter out of the mists of Madison Square Park in New York. Its silhouette was familiar and yet its effect was especially eerie. This was Maya Lin's Ghost Forest, a public art installation composed of forty-nine dead Atlantic white cedars, hauled in from the Pine Barrens in New Jersey and secured to the ground to mimic the appearance of a forest stand. Lin spoke of her desire to raise awareness of the nefarious effects of climate change, one of which being the saltwater inundation from a flooding estuary that kills the Jersey cedars. The trees were placed there to gray out and pronounced officially dead some months later. Their final moments of suffering, seasoned with a custom-made soundtrack of extinct birdsong, were meant to provide an ultimate purifying spectacle for the liberal folk of Midtown and *Frieze* magazine.[1]

Using trees as peons in environmental rhetoric is nothing new in the art world. Back in 1982, Joseph Beuys planted seven thousand oaks throughout the city of Kassel as an ecological commentary on technological society. Even before that, Alan Sonfist rewilded a block not very far from Madison Square Park with native species to create his Time Landscape. But Lin's brooding installation, perhaps shaped in response to the loud anti-scientific clutter of the Trump era, took a far more pessimistic approach toward the fate of the planet. If Sonfist's biome, which is discussed in a later chapter, was a rewind into an ideal American past, then Lin's forest was a projectile straight into a tree zombie apocalypse that may be just around the corner. But what makes Ghost Forest so uncanny is not only the grotesque treatment of the trees but their sudden appearance onsite. Most people would posit that groups of trees are not supposed to be moving around: they can be brought in as seedlings and planted to shade a sidewalk or decorate a backyard, but once this is done, they should stay put for a very long time. However, this intuition is not based on fact and is perhaps more indicative of a human need for stability than of the forests'

DOI: 10.4324/9781003473411-4

own immobile nature. While individual trees mostly move with their root tips that spread out tentacle-like in the invisible kingdom of the underground, entire forests do migrate constantly in response to environmental cues. A 2018 research based on Forest Service data found that 73 percent of tree species in the eastern US traveled west and north many miles and across counties.[2] Pine forests in the Himalayas were found to be climbing 19 meters uphill per decade due to changing climate conditions.[3] Studies of fossil pollen data show that tree migration was the norm rather than an exception throughout the Holocene. Research into the changing patterns of specific species demonstrates a dramatic increase in "range shift" over the past three decades.[4] Ample scientific evidence suggests that further forest migrations are surely yet to come.

Forest stability may be a fiction, but it is one that has been firmly rooted for millennia in the collective psyche of civilizations. In the eighth century BCE, the prophet Micah associated eternal peace with the image of men sitting under their fig trees. Cicero, quoting second-century BCE poet Caecilius, paired tree planting with posterity when writing that "he plants trees that they may be a profit to another age".[5] Around the same time in ancient India, Kauṭilya's master treatise on statecraft, *Arthásāstra*, linked large-scale forest planting with a stable source of income.[6] Various proverbs with obscure origins make that same point when connecting planting with past wisdom and future benefits. Prosperity is key, as the stable forest metaphor weaves together the long-term growth of humans and plants with notions of origin, permanence, and identity. Throughout different times, forest stability was cultivated by economic, social, or ecological arguments. However, it was always concerned with the clear demarcation and well-being of a community vis-a-vis the outside world, often described as hostile and perilous. For the dream of the long-standing forest is also the fantasy of a society that could withstand the damages of time and resist external threats. This is why stability is often found in alliance with conservative, protectionist, and sometimes nationalist sentiments. These alliances seek to shelter communities through regional planning and spatial design that utilize form to anchor these communities in relation to forest sites. In other words, tracing this forest metaphor reveals moments of perceived threats to existing ways of life. This metaphor is frequently found nested within theories and systems that seek to introduce order in contexts of social and political disequilibrium.

Stability was lacking in Mecklenburg ca. 1810, as a vision of a national economy based on spatial organization was being put forth. The small Grand Duchy of Mecklenburg–Schwerin, a territory in current-day Germany bordering the Baltic Sea to the north, was navigating its way in the stormy waters of the Napoleonic Wars. The dissolution of the Holy Roman Empire prompted Duke Frederick Francis I to seek neutrality at first and then join the Confederation of the Rhine in 1808, thus becoming a client state of the French. He then attempted to resist the French army, almost lost his duchy to the Kingdom of Denmark, and was finally able to join the German Confederation after the Congress of Vienna and secure sovereignty. During this volatile time, Johann Heinrich von Thünen, an agriculturist trained under Lucas Staudinger and Albrecht Thaler, purchased a lot in Tellow with the intent of turning it into a model farm. In that, he was following his teachers as well as earlier examples set by du Monceau, although with more modest means.[7] The work was developed based on years of observation and data gathering on the farm, but at the same time, it was shaped by von Thünen's underlying ambition to connect Thaler's agronomic concepts with a solid economic rationale following Adam Smith, whom he regarded as his other great influence. Already during his studies, he made note of an abstract geography that could be envisioned in order to calculate and optimize agricultural market activity. The idea was to assume a uniform land with a diameter of 40 miles and one city at its center to which all agricultural products would be sold as a springboard for the organization of different classes of agricultural economic systems.[8] Von Thünen spent ten years at his farm developing and testing his theories until he set to write what would become his famous work, *The Isolated State in Relation to Agriculture and National Economics: Studies on the Influence of Grain Prices, the Richness of Soil and the Yields from Crop Farming*. As its title suggests, von Thünen was trying to portray a stable picture of a national agrarian economy, smoothly run by rigorous mathematical calculations. As the actual conditions of Mecklenburg were still largely medieval, serfdom only being formally abolished in 1819, this intellectual endeavor required a great deal of abstraction. And indeed, some of the striking power of the theory rests on assumptions that transform current-day realities into a hypothetical geography. Building on his early ideas, von Thünen describes the *isolated state* as an abstract plain with no distinct geographic features, at the center of which lies a single, large city. In this model, all the land is of equal fertility, and the only means of transportation is the wagon. The people of the city supply manufactured goods in return for raw produce from those inhabiting the countryside.[9] These assumptions, once joined with economic principles of marginal productivity and rent, result in a specific spatial distribution of agricultural products throughout the plain. Simply put, each product is grown according to the costs of its transportation to the city and the rent the farmer has to pay. From this follows that agricultural activity is organized in concentric rings around the city: in the areas nearest the city, dairy and vegetables are grown due to the short time in which they have to get to the market. Next, one would find managed forests that provide timber for construction and fuel. In the following ring, people cultivate and harvest grains that can be transported easily over long distances. The outermost ring is reserved for grazing, as farm animals can be walked to the market to be sold. Surrounding the last ring is a wilderness that separates the *isolated state* from the rest of the world. The size of the rings is determined solely by the cost of transportation.

The *isolated state* was immersed in contemporary discussions within economic theory.[10] However, its abstract plains and concentric land-use rings found their audience with the help of the evocative illustrations included in the book. These radical visualizations, which the author warns are somewhat incongruent with the theory, gathered much attention and ushered in the subfield of locational theory, which dwells at the outer regions of economic analysis and geography.[11] Popularized in Germany by prominent figures such as Alfred Weber, August Lösch, and Walter Christaller, locational theory crossed the boundary from descriptive to prescriptive as its proponents envisioned "an economic science that, more like architecture than like the history of architecture, creates rather than describes!"[12] Von Thünen's illustrations eventually evolved from circles to hexagons to networks in subsequent theories that sought and were indeed instrumental in redesigning entire regions in Europe and the United States.

But even without overplaying von Thünen's impact in hindsight, his second concentric ring – the forest section – deserves scrutiny, as it postulates a different angle regarding the inclusion of forests within economic systems. Economic forestry in German states has several sources, which originated in the cameral class of Prussia and Saxony that, during the late eighteenth and early nineteenth centuries, introduced systematic surveying and stock calculation into the burgeoning field of forest management. Many of the forerunners gravitated toward the orbit of Heinrich Cotta and the forestry school he founded in Tharandt, and others further developed the mathematical side of the profession until German forestry became almost synonymous with the concept of soil-rent theory.[13] It is uncertain exactly how familiar von Thünen was with the works of his contemporaries. Neither the *isolated state* nor his many letters mention any such sources. Clearly, he was no forester, as he frankly admits when introducing the forest section: "For the foregoing calculation the author was not able (as he had been in discussing farming costs and product) to take his basic data from experience, but had to make a rough estimate as to the basic figures."[14] It was perhaps this lack of practical experience that allowed him to develop a unique perspective of forest management, which shared sources and insights with works of economic foresters yet diverged from them in some important aspects. The radical nature of the *isolated state* theorem simplified some problems raised by economic analyses of real-world forests as it assumed that "the primeval forests have vanished long ago, each wood and forest is man-made".[15] The notion of the total artificiality of forests purged economic conundrums such as mixed-age forests and was shared by later attempts to determine the most profitable time for harvest. The most influential of such attempts was Martin Faustmann's formulas from 1849, which assumed a "stylized plantation … [that] was a main vehicle in developing the theory of capital".[16] But whereas Faustmann's efforts were centered on taking snapshots of forests in order to relocate them to the realm of mathematics, von Thünen's calculations portrayed a complex and dynamic forest, expressed through clear spatial organization.

For the concentric forest ring was understood from the outset not as a fixed form but as one that is able to change both in total area and internal organization while retaining its ring logic. These changes would occur in response to economic fluctuations in costs or demand. Consider how von Thünen talks about changes in the outer boundary of forest: "The outer fringe of the forestry ring, no longer required for fuel production, will now turn to grain."[17] Elsewhere, he goes to articulate the inner subdivision of the forest: "Inside the forestry ring we would therefore find further subdivisions or concentric rings, producing different grades of timber."[18] These grades are firmly anchored in location:

> it may be profitable, in the innermost fringe of the ring of forestry nearest the Town, to produce trees with a very fast growth rate … the outer fringe, on the other hand, cannot afford to supply the town with any but the most expensive grade of fuel.[19]

These words expose a disparity between the circular diagrams accompanying the book and what the author had in mind. While the graphics convey a static image of spatial organization based on fixed distances from market, von Thünen paints an incredibly resilient forest form that is able to grow, shrink, or diversify in response to its economic environment (Figure 3.1).

The links between forests and the larger economic context are also exceptional in the *isolated state*. As the mainstream of economic forestry focused its attention on the question of maximum profit, that is, when

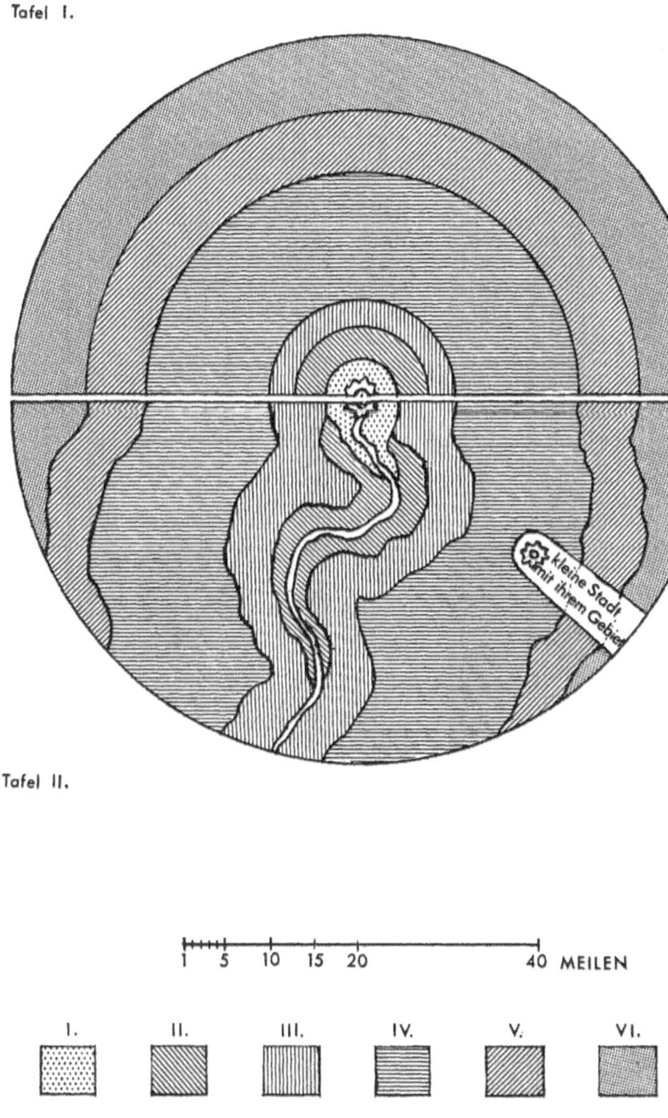

Tafel I.

Tafel II.

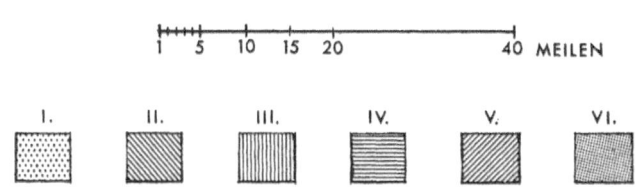

FIGURE 3.1   Plates of concentric land use rings from von Thünen's *Isolated State*, 1826.

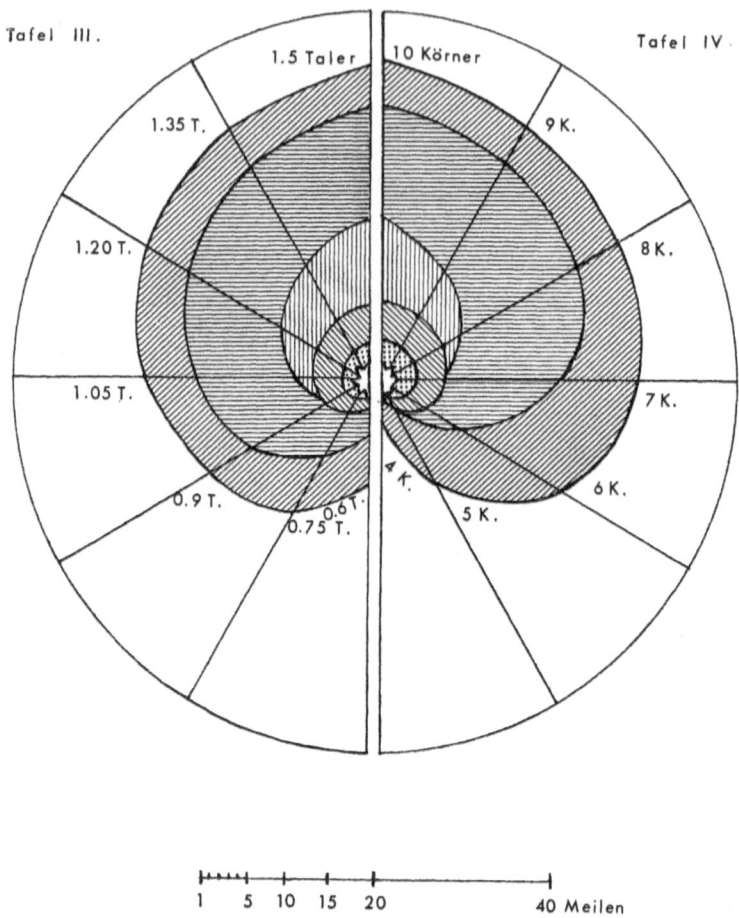

Tafel III.

1.5 Taler
10 Körner

Tafel IV

1.35 T.
9 K.

1.20 T.
8 K.

1.05 T.
7 K.

0.9 T.
0.6 T.
0.75 T.

6 K.
5 K.
4 K.

1   5   10   15   20   40 Meilen

FIGURE 3.1   (Continued).

would be the best time to cut down the trees in a given stand, its connections to larger contexts became frail. Faustmann's formulas, for instance, could only be linked to the general economy through the prevailing rate of interest.[20] In von Thünen's work, the forest is still analyzed and determined via mathematical calculations but is intimately linked to other assets and to the national economy by nature of location and proximity to its constituent parts. In lieu of microeconomic forest decisions, one finds an integrative macroeconomic system in which physical forms circumscribe but do not confine the workings of economic transactions. Despite what might be expected, it was the austere abstraction of real-world conditions upon which the *isolated state* was built that allowed a renegade idea of national stability to emerge.[21] The model's innate interchangeability of land use and open-endedness regarding its spatial organization set it apart from classical economic theories and their focus on equilibrium. In what was described by prolific urbanist and planner Peter Hall as the world's first economic model, the concentric forest form was included for its ability to mediate complex pressures and adapt to changing conditions.[22]

The cardboard history of forestry in the German states and in unified Germany tends to conflate scientific management with economic planning to present a beacon of rationality capable of analyzing and transforming forest landscapes.[23] This is partly due to the enormous success throughout the colonial world of German foresters such as Dietrich Brandis, who reached a zenith of celebrity and influence around the turn of the century. However, in real time, forest economics and science were often incongruent. Mathematical analyses of high abstraction such as Faustmann's were regarded as impractical on the ground, and conflicted with the idea of sustained yield, which "had become the central idea of forestry in Germany in the nineteenth century".[24] As opposed to the soil-rent theory, which sought to determine the optimal time for felling in order to maximize profit, sustained yield developed out of the ambition to extract revenue for perpetuity, based on natural tree growth cycles and other forest characteristics. This principle, allied with the idea of political and social stability, became most effective as it was easier to explain and more appealing to both colonial administrations and federal agencies. When, in 1893, Gifford Pinchot published a report on the "first practical application of forest management in the United States" in Biltmore Estate, he followed this principle and presented a working plan that was meant to allow "a nearly constant annual yield" for over 150 years into the future.[25]

As Pinchot went on to organize the nascent US Forest Service, sustained yield took a back seat in the quest for bringing more lands under federal control. During that time, *scientific forestry* was used as a blanket term in the philosophy of an agency that proclaimed itself as advancing the public good. This zeal was often contested by operators big and small, who regarded the agency's "Pinchotism" as expansionist and anti-business.[26] Furthermore, while the principles of sustained yield were taught in most forestry schools in the country, its applications were few, as American forests did not face the same issues as their European counterparts. Most national forests, containing old growth, did not conform to the even-aged model assumed by European sustained yield, and so the Forest Service developed policies that focused on regulation that would gradually transform its forestlands into regulated forest structures.[27] From a market perspective, dwindling supply that haunted Europe for centuries was hardly an issue. In fact, the American market of the early twentieth century was besieged by overproduction. Thus, industrialists had no incentive to engage in the long-term management of their lands: it was easier and more cost-effective to cut the lumber and let the land revert back to the county in order to avoid taxation.

★★★

This changed after World War I, as "timber famine" was being discussed more frequently in professional circles. During the 1920s, the Forest Service began to advocate for federal regulation of cutting in both public and private lands, an idea bitterly rejected by the private sector until the economy came to a screeching halt in 1929. It was only then that forestland owners were hard-pressed by dire prospects to consider forms of collaboration with the government. Sustained yield was brought up as the right concept that may just save the day. Largely responsible for this introduction was David T. Mason, a Yale-trained forester who gathered Forest Service experience in both western forests and Washington, D.C., and aligned himself with industrial leaders through his number crunching capacities and expertise in both the monetary and scientific aspects of forest management.[28] As early as 1920, Mason entertained the idea of "continued yield" as a value proposition that could appeal to timberland owners.[29] By 1927, this intuition matured into a reinvention of sustained yield as a holistic approach that could answer to scientific, commercial, and social challenges. His envisioned sustained yield management unit included "not only the forest land involved, but also the logging development, the mills for conversion, and the community economically dependent upon the enterprise".[30] With that, he expanded sustained yield beyond the physical boundaries of forests to include the entire working cycle, a locational economic concept that resonated von Thünen's ideas. In the context of a crisis-struck economy, the stabilization of communities became a priority in Mason's work. In a publication on the subject, he opened by presenting a diagram titled "Forest Conservation Problems and their Solutions" in which the problem of stabilizing local communities was listed first, above national supply and profitable operations.[31] "Such communities", he writes "can be stabilized *only* through sustained yield forest management."[32] However, this change of focus demanded a high level of cooperation between private and public forest management in order to regulate cutting over the long term, which was easier said than done. Mason spent much time in the following years to advocate for the idea and found some reception with industry leaders such as the Weyerhaeuser company, which saw in Roosevelt's presidency the mark of an era of "controlled individualism".[33] The company also hired Mason to make a sustained yield study in its St. Helen's unit and adopted its recommendations.[34]

The federal administration followed with the creation of the Lumber Code Authority under Roosevelt's National Recovery Administration, a process in which Mason was closely involved.[35] However, the pinnacle of his sustained yield efforts was the drafting of the Sustained Yield Forest Management Act of 1944, which brought into existence far-reaching experiments in both forest and community design. The act authorized the secretary of agriculture and the secretary of the interior to enter long-term agreements with private owners to establish "cooperative sustained-yield units" of two types, giving companies exclusive access to federal timber in exchange for the adoption of sustained yield practices.[36] The first declared goal of the units was to "to stabilize communities, forest industries, employment, and taxable forest wealth".[37] In other words, the act envisioned spatial units that were insulated from volatile market environments through the allowance of monopolistic conditions – small *isolated states* in the midst of a raging sea of instability. Theoretical as it may sound, the proposition proved very attractive to forest companies, and seventy-six applications for new units were submitted to the Forest Service. The evolutionary process of sustained yield units, only one of which was established as a cooperative unit and five others as federal units, disclosed the vehement opposition by those outside the circle: communities, companies, and landowners who felt disadvantaged by the proposed arrangement. Attacking the idea as monopolistic, socialistic, and unconstitutional, they managed to eventually curtail the unrolling of new units and laid bare the limitations of federal schemes of social engineering.[38] However, the fate of the entire program does not take away from the merit of realized units, especially in the case of the more radical sustained yield experiment in Shelton, Washington (Figure 3.2).[39]

The 270,000-acre unit, immediately west of Puget Sound, came into existence with the signing of a hundred-year agreement between the Forest Service and the Simpson Logging Company. Simpson, the largest landowner in the region, had been preparing for this moment for over a decade, willingly adopting practices of advanced forest management and prompting the Forest Service to author feasibility reports on its potential to implement a program of sustained yield in Mason and Eastern Grays Harbor counties.[40] These reports left no doubts regarding the urgency of taking action, showing the absolute dependence of the region on the forest industry and arguing that all merchantable timber within the working circle would be depleted within a decade. While the precise boundaries of the unit were debated, it was clear from the outset that it would include

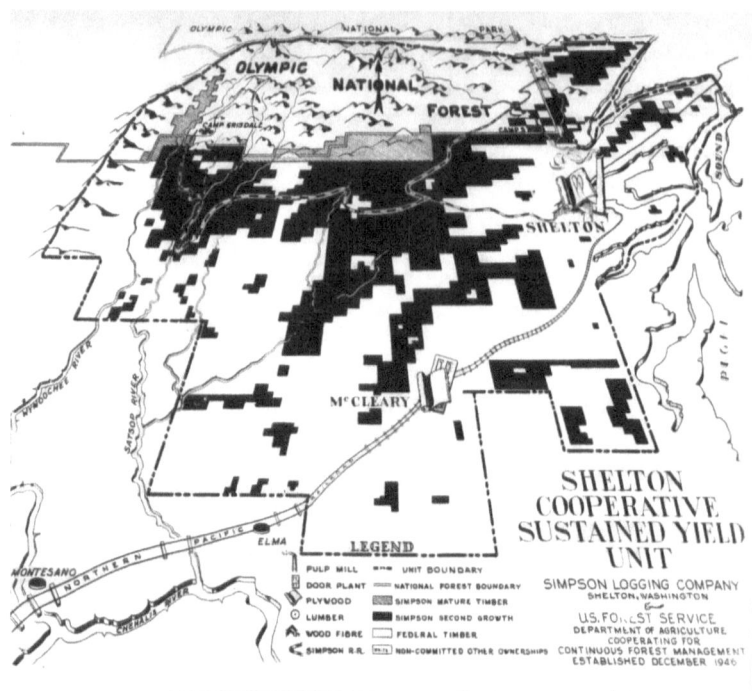

FIGURE 3.2   Plan of the Shelton cooperative unit, 1954. Courtesy of Harvard
Libraries.

parts of the Olympic National Forest, increasing the company's timber
pool. In a return to the boon, Simpson committed not only to a better
utilization of forests but also to the sound management of socioeconomic
aspects in the region, including "higher local employment, manufactur-
ing capacity matched to the available resources, and an end to fluctuations
in population and payrolls".[41] In presentations to the public, the Forest
Service maintained that in giving form to working circles, essential links
would be established between the location of milling or logging communities and that of the timber itself, "so that woodworkers will have
an opportunity to live at home in permanent communities and to com-
mute to and from work".[42] At the Shelton Unit, at least, these optimistic
prospects were partly fulfilled. Various accounts written in the following
decades described the impacts of island stability: investment in new plants,
plantation of tree farms, more financial institutions and more bank depos-
its, and more citizens cheering the hundred-year contract.[43]

The massive spade-shaped form, designed to pin down resources,
money, and people, did manage to do so comparatively well. At the same

time, it prompted organized resistance that disabled further attempts by the Forest Service to establish other cooperative units. The federal sustained yield units that were eventually established, supposedly simpler in terms of organization, were soon marred as local interests and antagonisms complicated their chances of success until most were abandoned.[44] The Shelton Unit did not reach the hundred years mark, but it did survive until 2002, when it was terminated by mutual agreement. And while Shelton remains one of a kind in the history of federal forest-derived regional planning, its relative success may have something to do with the specific land history of the region. That is, since the designation of its form was also an attempt to rectify a history of land ownership in the Pacific Northwest that intermingled private and public lands thus putting them at an inherent conflict of interest. Aligning these interests through economic and social engineering was a way of defragmenting a shattered territory that was inherited from the original sin of transferring large tracts of land into private hands in the mid-nineteenth century, without checks regarding the health and longevity of either forest or people.

While it may be considered like an end of a process, the proposition put forth in Shelton was in fact anticipating other stabilization efforts based on forests, which utilized form as a means of achieving greater social and, later, environmental and ecological equilibrium. Consequently, the design of forests received a thrust that made it the concern not only of professional foresters but also of scientists and regional planners seeking to counter instability in its various manifestations.

***

Linking the deliberate design of forest form with the achievement of stability acquired new meanings with the development of ecological perspectives on natural systems, which, starting in the 1960s, permeated forest science and forest planning. Understanding the forest as a dense web of complex interactions between given structural features, species composition, and external disturbances meant that the idea of stability should be somewhat modified to include the notion of systems in a state of dynamic equilibrium.

In 1955, ecologist Robert MacArthur published a text on community stability. Focusing on animal populations, he proposed a theoretical model that considered their food webs as energy flow diagrams. This abstraction allowed him to borrow from probability and communication theories and pose that under certain assumptions, species populations approach constant values, thus reaching stability.[45] Well aware that the assumptions that lead to the neat first diagram are impossible in real systems, he moved on to present a second one that accounts for energy losses in a given food web. He then established stability as an area of probability, with maximum and minimum values for each combination of species, thus factoring it in as a calculable metric, related to the number of links within food webs and the survival rates of species in a specific environment. The proposition came with a disclaimer about the nature of stability: "a difficulty arises", he writes, "in making this definition quantitative, because our intuition of what stability means is ambiguous … [the] choice among various functions to define stability precisely rests only upon usefulness of the definition"[46]; in other words, here, too, stability was a question of defining who and what is in the circle and who and what is left outside.

In 1961, MacArthur ventured to bridge the gap between ecology and taxonomy and outline a theory of species equilibrium. Echoing Darwin, he and Edward Wilson focused on "insular biogeography" – island conditions in which the relationships between the immigration, adaptation, and extinction dynamics of populations and the physical features of island conditions could be described and theorized. Right at the outset, the work established a correlation between island size and the number of species and, more significantly, the difference in correlation in isolated and non-isolated contexts.[47] At the same time, the authors argued that insular habitats were not restricted to actual islands but are likely to occur wherever development meets the natural world. Using maps of the increasing fragmentation of forest area in Wisconsin as illustration, they write: "The same [theoretical] principles apply, and will apply to an accelerating extent in the future, to formerly continuous natural habitats now being broken up by the encroachment of civilization."[48] With that, the work helped establish the idea of heterogeneity in both time and space within ecological theory. In other words, it meant that the physical properties of a place determined its ecological development and were ingrained in the patterns of dynamic equilibrium that could emerge: a crucial link between form and ecological performance was thus created.

In the Pacific Northwest, forest fragmentation was a reality for a century. This was due to the intensive logging that was taking place since the white man laid eye on the cascades but even more so due to the curious history through which land was transferred to the possession of private companies. As mentioned elsewhere in this work, the federal government granted railroad companies a substantial amount of land, at times reaching a 40-mile strip for each mile of laid track, to sponsor their costly operations through land sale and resource extraction. The land grant followed the logic of the 1785 Land Ordinance in dividing the territories into 1-square-mile tracts, but land was handed over in alternating sections, leaving federal lands in between. On one hand, this manner of provision was a matter of speculation, as the government was hoping to sell its lands as the rail would increase their value. On the other, it demonstrated a hesitation in providing the most powerful private corporations at the time with uninterrupted tracts of land in unincorporated territories. The result was a checkerboard pattern, visible on the ownership maps that accompanied the grants, that became a widespread territorial pattern as railroad companies cut over and sold their lands.[49] Since clear cutting was a standard practice and regrowing trees on cutover lands took many decades if it happened at all, the pattern became characteristic of large areas in Montana, Idaho, Washington, and California, colloquially known as the Chackerboard Cascades. As ecological theories of insular biota and population dynamics gradually made their way into forestry, it was becoming clear that prevailing fragmentation – the inherited form of the region's forests – established ecological patterns that were detrimental to forests and their communities (Figure 3.3).

FIGURE 3.3   Aerial photograph of the Checkerboard Cascades in Montana, 2008. © Google Earth.

At the same time a new interest in old-growth forests, mostly found on federal lands, emerged. These stands, containing trees hundreds of years old and mostly past their industrial prime, were beginning to be regarded as islands of complexity and potential. Surprisingly, field research on old growth was almost nonexistent. The lack of commercial interest and the long-standing focus of forest scientists and managers on yield metrics kept it out of the limelight, and well into the 1970s, it was regarded by foresters and the academic community as "over-mature" or "decadent".[50] This negligence prompted plant ecologist and forester Jerry Franklin to organize a group of scientists and land managers who would work on identifying the ecological characteristics of western old-growth forests. Working at the time as a research scientist for the US Forest Service, he tapped into the funding of the International Biological Program in 1969, which was provided by the National Science Foundation and other federal agencies.[51] This program aimed to study various "typical" ecosystems and included the western coniferous forests as one. The small group of scientists from Oregon State University and the Forest Service initially set out to study old growth at least to "have a historical scientific study of what it was".[52] Focusing on a watershed at the H.J. Andrews Experimental Forest in Oregon, the team soon realized that what they found differed significantly from accepted forestry views. The carbon budget showed substantial amount of dead wood on the forest floor and in the streams, and it was clear that within that ecosystem, "dead wood is not really dead … [but] full of all kinds of living things".[53] As work went on in the field, Franklin became convinced that the study of old-growth systems not only presented science with new avenues for research but could also inform new management regimes for planned forests, enhancing their endurance and stability.

The group then set out to compare old-growth and younger forests where data were available, particularly targeting the managed (young) conifer forest model as it was defined in the protocols of the US Forest Service. This comparative methodology lent the group's analyses a thrust that went beyond just science, underwriting a different way of thinking about and practicing forestry. The report, commissioned by the Forest Service and published in early 1981, articulated the old-growth forest as a complex wonderland. Diagramming the fauna along streams fed by decomposing logs or outlining divergent shapes of individual Douglas fir trees, the foresters sought to scientifically describe and learn from environments that were hitherto considered mostly through commercial avarice or transcendentalist fascination.[54] With that, the report not only proposed to focus attention on the question of better forest management but also implicitly sketched out an antithesis to prevalent ecological abstractions as packaged, for instance, in Macarthur's theories. It was an opportunity for western forest scientists and managers to strike back at a scientific discourse that they understood as somewhat detached and uninterested in the realities of American forests.[55] The team extended their research into canopy studies, which turned out to be tremendous productive systems, and the notion of accumulated carbon storage began to be better understood. This intense research and its forest management conclusions not only circulated within the scientific community but began to influence forestry practices on both public and private lands. In many ways, it was responsible for a transformation of the idea of forests and with it the notion of stability. Rather than focusing on regulating the long-term extraction of wood material, the new approach focused on the carbon and energy budget within a forest ecosystem. And so, as Franklin puts it: "It turns out that old-growth are not cellulose cemeteries ... or biological deserts. They are not falling apart but are very stable systems in terms of carbon" (Figure 3.4).[56]

In 1985, Franklin took a pause from intensive fieldwork to go to Harvard Forest, where he collaborated with Richard T. T. Forman to introduce form as an operator under this new definition of forest stability. Interested in possible correlations between Forest Service cutting practices and catastrophic ecological disturbances, Franklin proposed to study the consequences of the pattern by which one imposes harvest units on a virgin forest.[57] Bringing this practical question in contact with Forman's ideas about landscape ecology, which hypothesized connections between landscape patterns and ecological performance, the two began a process in which they designed a "pretend checkerboard landscape" – 1,000 hectares divided up into little 10-hectare blocks – and studied the consequences of dispersing the harvest units over long period. Analyzing fire, wind, and disease factors, they quickly realized that such a cutting pattern maximizes ecological fragmentation. The patch size and the form and distribution of cut areas that the Forest Service was doing made sense from a managerial point of view, but it created a very vulnerable edge, which was disastrous from an ecological point of view.[58] But something else happened in the process: as Franklin worked his way through the imagined checkerboards, the analytic function of formal analysis was followed by the idea that form could be utilized to improve ecological functioning (Figure 3.5). Upon returning to the Northwest, he convinced the Forest Service to try and aggregate cutting, which was partially adopted until old-growth cutting was forbidden altogether soon after.[59] In his new position at the University of Washington, Franklin got involved more and more in policy,

Tree density in age series of stands

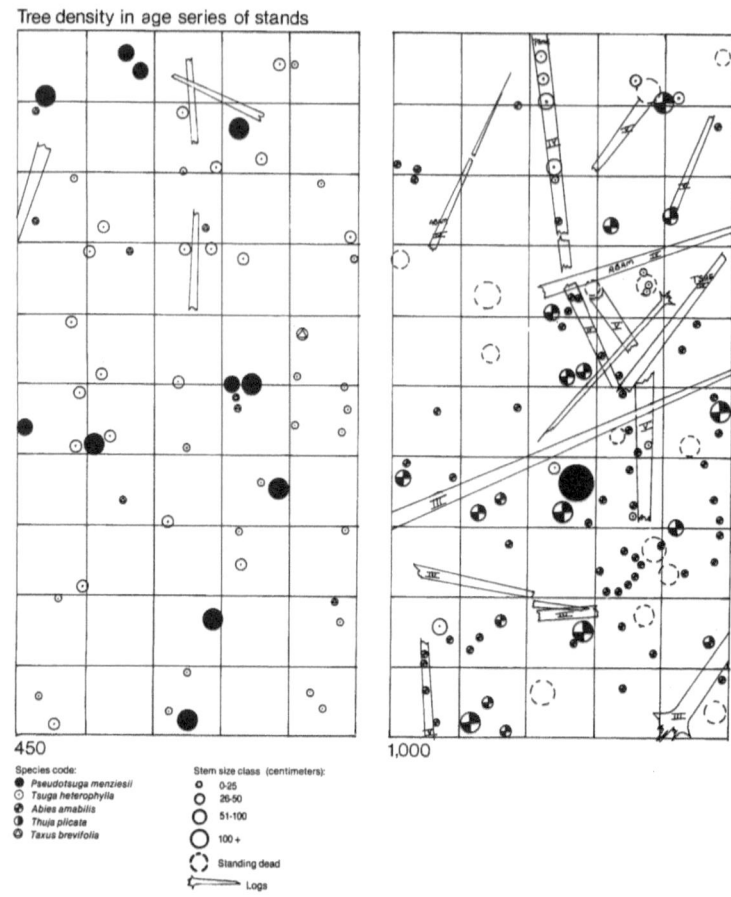

450   1,000

Species code:
● Pseudotsuga menziesii
⊙ Tsuga heterophylla
❷ Abies amabilis
❸ Thuja plicata
❹ Taxus brevifolia

Stem size class (centimeters):
○ 0-25
○ 26-50
○ 51-100
○ 100 +
⊙ Standing dead
⟋ Logs

17

FIGURE 3.4   Diagrams of old-growth forests from *Ecological Characteristics of Old-Growth Douglas-Fir Forests*, Franklin et al., 1981. © USFS.

cumulating in the Northwest Forest Plan, adopted in 1994, which was designed to administer an area of 10 million hectares based largely on scientific principles.[60] Regardless of its shortcomings, the plan did much to establish the status of form and pattern as essential elements in large-scale ecological design. Moreover, it galvanized a new image of old-growth forests as havens of stability, called at one point the "Fort Knox of carbon" – an image that has become consensus today.[61]

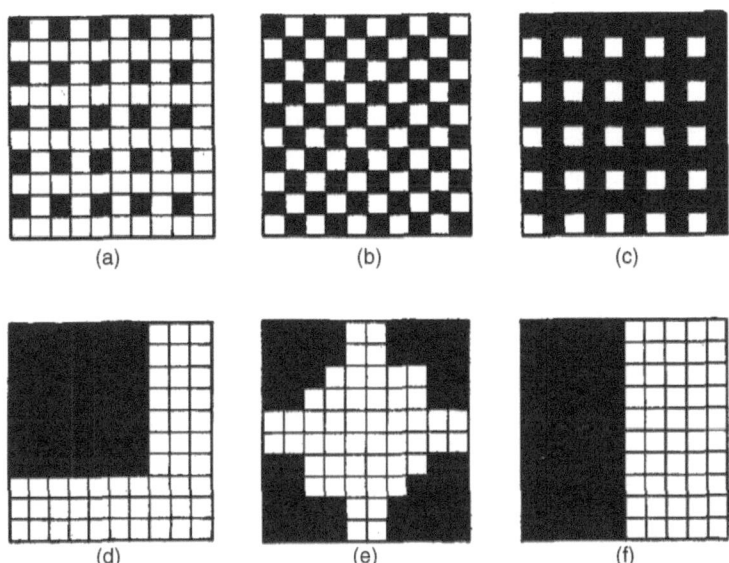

Fig. 2. Patterns of clearcutting developed under various models. (A-C) Progression of clearcutting using the dispersed patch model in which areas are selected for cutting so as to be distributed regularly across the landscape; shown are 25, 50, and 75% cutover points. (D-E) Pattern of cutting at 50% point using single-nucleus, four-nucleus, and progressive-parallel cutting systems.

FIGURE 3.5   Diagrams of hypothetical clear-cutting patterns included in Franklin and Forman's article, *Creating Landscape Patterns by Forest Cutting*, 1987. Courtesy of Jerry Franklin.

★★★

The 1990s were also a time in which architects began to experiment with the potential of form to initiate and channel ecological processes. Two such projects are the unrealized proposal for the Hannover Expo masterplan by Herzog & deMeuron (1992), and the Office for Metropolitan Architecture and Bruce Mau's winning proposal for the redesign of Downsview Park in Toronto (2000). While the two projects remain somewhat idiosyncratic within the portfolio of each firm, they demonstrate the extent to which forest form occupied the imaginations of major players in the architectural field during that time.

The idea for Expo 2000 was initiated in the late 1980s by Hannover's influential mayor Herbert Schmalstieg, who teamed up with Birgit Breuel, then minister of transport and economy in Lower Saxony, and applied to the Bureau International des Expositions, which elected Germany as the host country for the international exposition. Taking place in a recently reunited country and under a flag of liberal optimism, Expo 2000 sought to implement the United Nations Agenda 21 for sustainable development.[62] The city of Hannover was hoping to benefit from public and private investments to revitalize its existing trade fairgrounds, build a new housing area, and modernize transportation infrastructure. In 1992, the city council and the state of Lower Saxony announced an international planning competition for an area that included the fairgrounds and the adjacent Kronsberg area, a patchwork of industrial agriculture and rural settlements. The competition documents specifically called for the consideration of sustainable development by requiring applications by multidisciplinary teams that were to include at least an urban planner, an architect, and a landscape architect. Additionally, an environmental impact analysis (EIA) was set up in 1990 to ascertain the environmental consequences of the Expo on the Hannover region on an ongoing basis. Competing teams were expected to respond to the initial findings of the EIA and demonstrate previous experience in working with process-based evaluations. Beyond the more common parts of the brief, the competition organizers emphasized the creation and enhancement of ecologically stable, semi-natural countryside areas included in the site.[63]

In its submission, Herzog & deMeuron responded to these imperatives by proposing a vast planted forest in which both plants and buildings were designed under the same logic. The new plan, which they named Kronsberg Forest, took its outline from the topography of a nearby hill and the arching contours of surrounding highways. However, its internal organization was determined by defining close relationships between building and forest architectures. The firm invested much effort in historicizing this relationship, looking at previous architectural types and carefully defining their appropriate forest counterpart. In the competition text, the firm wrote:

> For their part, the buildings take the trees into account. These are low buildings that, as do Bruno Taut's Uncle Tom's cabins, hide in the woods or tall buildings which rise above the woods like Le Corbusier's Unités or flat installations that seem to meld together with the tree-tops.

In searching for precedents, Herzog & deMeuron was looking for cases in which trees were not regarded as addendums to buildings but were essential for the success of specific architectural concepts. By doing that, it also attempted to link the histories of forestry and architecture, establishing a case for a deeper symbiosis that could exist between the two (Figure 3.6).

The physical maquette produced for the proposal is a reincarnation of forest models used by governmental agencies to promote ideas of sound forest management that stabilizes forested areas or entire regions.[64] It is composed of countless identical trees that are distributed on the topography, forming a clear image of neat organization and aesthetic tidiness. Interestingly, while coming from a firm that made a name for itself by focusing on clear geometric massing and meticulous building details, it is hard at first glance to even see the buildings in the model. Upon closer inquiry, it becomes clear that the reason for this is the fact that Herzog & deMeuron attempted to apply the same underlying grids to organize both trees and structures. In other words, the firm attempted to use form to organize the two components as part of one forest system.

There is a certain ambivalence built into this proposal. Clearly, the entire scheme is an artifice: the forest would have to be planted and many of the buildings be built from scratch. But the firm was aiming to explore a course of action that would utilize the umbrella of sustainable development and ecological stability to save architecture from its own hubris. Jacques Herzog, reflecting on the project, said that the forest could "relieve us as designers from always producing these objects … as [the forest] creates something much bigger and powerful as an environment, the buildings don't have the same prominent role they now have in our cities".[65] In time, it was hoped by the architects, such a way of building could become the blueprint for the city of the future. The project lost the competition to a less experimental scheme of a sustainable urban district, and Herzog lamented the fact that its scientific ecological aspects could not be further developed. Nevertheless, Kronsberg Forest remains valid in its conscious attempt to emphasize the capacity of designed form to determine ecological and human habitats in some manner of coexistence. In retrospect, the proposal, curiously set in a region where forestry emerged as a scientific managerial discipline two hundred years prior, can also be regarded as presenting a parallel trajectory to the tendency to hide architecture and simply drown cities in green. It remains an antidote to what Herzog calls "a more decorative attitude … where you build trees into buildings".[66]

The OMA Downsview Park project was more blatant in its use of form to argue for ecological and social impact. The Toronto site, a 120-hectare decommissioned military base, was designated in 1998 by the federal government as "Canada's first national urban park". An open international

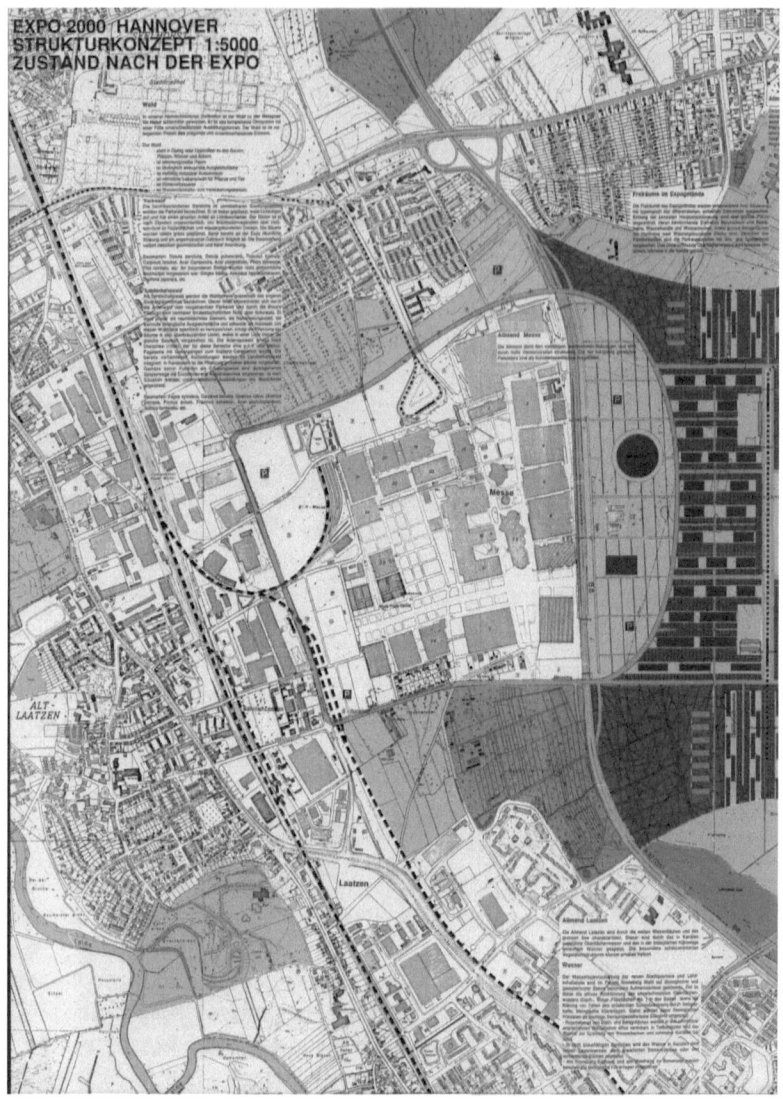

FIGURE 3.6   Hannover Expo 2000 competition entry, Herzog & deMeuron,
1992. © Herzog de Meuron.

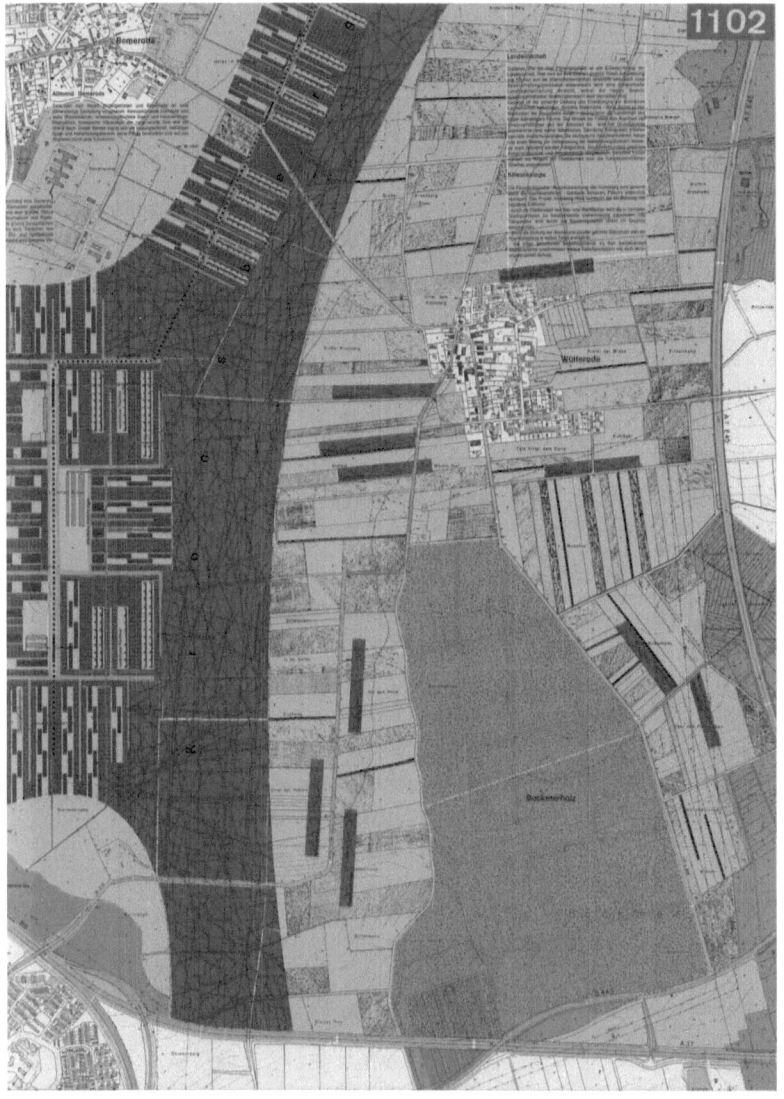

design competition was announced the following year, attracting 179 submissions from around the world, out of which five proposals were shortlisted and advanced to a second stage. OMA's winning scheme, titled Tree City, was developed with Toronto-based architects Oleson Worland and prominent Canadian graphic designer Bruce Mau. The proposal outlined a matrix of circular tree clusters, covering approximately 25 percent of the site, and interconnected through a dense network of cycling and walking paths. The project envisioned a soil remediation process, in which the site will be seeded with clover and then with wheat and barley, to prepare it for future landscape design interventions. The proposal captured the imagination of the jury because instead of planning a park, it attempted to use form in order to create an infrastructure that would be stable enough to sustain the financial and political uncertainties marring the feasibility of the project.[67] Highly rhetorical and highly graphic in approach, the project deliberately abstracted the site to a flat surface covered by circles, and garnished the proposal with equations that seductively interwove ecological and social values: "Grow the Park + Manufacture Nature + Curate Culture + 1000 Pathways + Destination and Dispersal + Sacrifice and Save = low density metropolitan life."

As obscure as it may appear in hindsight, this recipe approach allowed for the creation of a scheme in which future unknowns could be accommodated without compromising the core principles of the project. Curiously, while trying to bypass design in the traditional sense, planning for unforeseeable futures was anchored in circular forms.[68] It is tempting, even if somewhat futile, to consider the ecologies that would be designed by this total formal strategy had the park been materialized according to the proposed scheme.[69] Clearly, the deliberate fragmentation of landscape patterns and the superfluous road networks would maximize edge conditions and cut off biotic interactions to the point of rendering the parc almost ecologically meaningless. While the programmatic strategy was hailed as radical and groundbreaking, when considered from a forest standpoint it is a conservative monocultural scheme where animal and plant life are constrained within circular cages. Perhaps more than the forward-looking sagacity identified by the jury, OMA's Downsview scheme represents form's potential, and eventual failure, in achieving stability in a context of inherently unstable environments.

***

Far away from Toronto, in the very different context of the Ethiopian highlands, a more nuanced approach to forest circles had been developed. The sites of these circles are the church forests of the Ethiopian Orthodox

Tewahedo church, which are beginning to be regarded as ad hoc ecological stabilizers in a largely deforested region. The complete history of the formation and distribution of groves surrounding churches in the central and northern highlands is still in development. However, there is a growing acknowledgment of their functioning on both the local and regional levels. A 2016 study identified through sample satellite images 394 such forests, which were 2 hectares in size on average and separated from the next forest by at least 2 kilometers. It then used these figures to calculate a total number of 19,400 church forests throughout the highlands, covering an area of 39,000–57,000 hectares.[70]

In itself, a forest surrounding or containing a holy place is a common phenomenon, associated with various religions in various parts of the world. Sacred groves have a long-standing presence in non-Western and pre-Christian religious systems, from the oracular oak grove of Dodona in the second millennium BCE to Chinju-no-mori in ancient Shinto, which likely preceded temple architecture, and to the Gaul forest sanctuaries attacked by Julius Caesar. In many cases, from the village groves in Tamil Nadu to the ones maintained by the Sani people in the Shilin Yi Autonomous County in Yunnan Province, these stands were created and survived through the ages by the stewardship of local communities. In Africa, several such areas, like the Osun-Osogbo site in southern Nigeria or in the Pare mountains of Tanzania, were considered as rarified exemplars of primary forests that gave way to development and agricultural expansion over the last century. However, this view is beginning to be contested by ecologists and anthropologists, giving way to a complex understanding of sacred groves as purposefully designed landscape elements where social, political, and cultural dynamics interact and adapt over time.[71]

The Ethiopian church forests share similarities with some of these other sacred groves, but they are strikingly different in the intimate connection between doctrine and spatial form. The origins of the Orthodox Tewahedo Church in Ethiopia are found the second quarter of the fourth century, when King Ezana and his court converted to Christianity. In the following centuries, it was amalgamated into a form of state religion with the ruling elites while gathering popular adherence among the farming communities of the Ethiopian highlands.[72] In the late medieval period, churches of simple circular and rectangular shapes began to be constructed in these areas. The internal structure of these churches is organized in three concentric rings, reflecting different degrees of sanctity. The innermost part – the *Maqdes* – contains the Tabot, which represents the Ark of the Covenant, and can only be approached by the priests. The second ring is the *Keddists*, reserved for communicants, and the third, the *Qene Mahelet*, is for the entire congregation. This purity gradient spills over to the church's courtyard or immediate surroundings. Sergew Hable Sellassie and Belaynesh Mikael note that "the church precincts and the surrounding wall are considered sacred, therefore those who remain outside the church during the service are considered to have attended church".[73] In the Ethiopian highlands, widely forested until the twentieth century, this meant trees were included in the holy area of influence.[74] As the landscape around them was deforested and quickly transformed into agricultural land, the church forests emerged as circular islands within a sea of monoculture, bringing attention to their unique ecological functioning, derived from their exceptional form and placing (Figure 3.7).

While the churches have existed for tens and sometimes hundreds of years and the degradation of the landscape has been going on for decades, attention from the scientific community is quite recent. In 2007, Alemayu Wassie Eshete submitted his PhD thesis on the subject at Wageningen University.[75] Wassie, a native of northern Ethiopia, spent years in his early life training to be an Orthodox priest before focusing his attention on scientific education. Prior to writing his doctoral thesis, he already published articles attempting to gather scientific information and portray an ecological picture of church forests. The PhD work went deeper into studying the structural and species composition of twenty-eight forests in various altitudes and conducting soil seed bank studies to analyze their regeneration prospects.[76] What was most significant in this work was the systemic understanding of forest churches as an archipelago of ecological potential that serves surrounding communities, and the proposition that their survival depends on active conservation and sustainable future management of forests. Wassie's thesis received a prize

FIGURE 3.7   Circular forest church of Weger St. George in the Ethiopian highlands, 2021. © Google Earth.

from the Association of Tropical Forestry and Conservation, through which he got into a conversation with American biologist and canopy expert Margaret Lowman.[77] The brilliance and urgency of Wassie's work were clearly evident, but he was struggling to find direct solutions that could ameliorate degradation and answer to the priests' immediate challenges in conserving their forest tracts.[78] Lowman organized for Wassie to spend time as a visiting researcher in New College, Florida, during which they used satellite imagery from Google Earth to validate their suspicions regarding the dramatic scope of deforestation and devise in initial form the idea of building conservation walls around the church forests. As they traveled to the region, satellite images were used to communicate to the priests the dire conditions of their forests, and the walls were offered as the most viable and effective solution to their conundrum. Wassie and Lowman managed to mobilize a group of religious leaders, and a year later Lowman witnessed one such drystone wall, 1.6-kilometer perimeter, safeguarding the biodiversity of one forest. She notes that it was a clever solution from both an ecological and social points of view, as it "excluded cattle and goats, defined the forest boundaries for farmers, minimized firewood collection around the edges, saved the seed bank and biodiversity, and created a great source of pride for the local people".[79] The walls were adopted by more congregations, and Wassie and Lowman went on to acquire funding for the building of more walls, each with four gates on four sides of the circle, and maintenance of both church and forest. This simple strategy was cost-effective as well, making it one of the "cheapest countrywide biodiversity conservation programs in the world".[80] As the strategy of responding to the ecological island pattern by connecting church forests for better resilience is still being debated by scientists, the proposition to better define the boundary of each may seem counterintuitive. However, research by Wassie and others not only supported the correlation between the exact forest form of a specific island and its biotic characteristics but also laid the foundations for the argument that by fixing their circular precincts, these islands' contribution to regional economic and ecological stability could be maintained and perhaps enhanced.[81]

***

While attempts to stabilize church forests as a regional economic and environmental strategy were only scantly supported by the Ethiopian government, it got involved in the lavish attempt to ecologically stabilize Sahelian Africa by planting a 7,000-kilometer-long, 15-kilometer-wide forested band stretching across the continent from Djibouti to Senegal.

The project was first proposed in 2005 and advocated by Chief Olusegun Obasanjo, former president of Nigeria, and by Abdoulaye Wade, president of Senegal. In 2007, it was adopted by the African Union and enshrined as the Great Green Wall. The initial graphics showed the wall as a shiny green ribbon, emphasizing its function as a barrier to the encroaching Sahara Desert. And while the actual project has developed since to include a more complex multi-goal approach to land management, the ribbon remains a useful media asset, instrumental in securing international support without which the Green Wall would wither and disintegrate.[82] The Great Green Wall of Africa, while specific in social and economic challenges, is modeled after previous continental green strips such as the Three-North Shelter Forest Program, active since 1978. Designed to halt the expansion of the Gobi Desert, the program was conceived as a series of large-scale wind-breaking tree stands, forming in final form an immense 4,500-kilometer-long belt. China boasts the planting of more than 500,000 square kilometers, representing 6 percent of its total land area, and making the initiative the largest artificial forest in existence. Large-scale afforestation as official policy began decades before the current program, as the Chinese Communist Party came to power. In the early 1950s, officials advanced the planting of a 1,100-kilometer shelterbelt in northwest China and called for afforestation of 50 percent of its barren hills within thirty years.[83] These declarations and the twelve-year afforestation plan that followed in 1956 were designed in response to severe timber famine and soil degradation. Under the national focus on industrialization, the plan emphasized fast-growing monocultural forests close to the points of demand.[84] The Chinese afforestation plans were influenced and largely modeled after the 1948 Soviet plan to create shelterbelts throughout the Russian steppe under the hypothesis that these actions would break the winds coming from Central Asia and prevent draughts. A well-known propaganda poster shows Joseph Stalin with a pipe in one hand and a green pencil on the other, designing vast meandering forest belts that connect the Caspian Sea to Uralsk and to Stalingrad and straight forest bands laid perpendicular to one another that divide the entire region.[85] While tree planting in the steppes was a decades-long Russian engagement, shaped by a blend of scientific and imperial ambitions, the Stalin-era shelterbelt program was in fact inspired by the American experiment in the windswept Great Plains.[86] Protective planting had been practiced by farmers in the Plains for several decades, and the Forest Service was looking at the utility of windbreaks since the early twentieth century, curiously citing the Russian steppes as an important antecedent.[87] However, in 1934, these experiments became official policy as the Roosevelt administration announced a proposal to plant shelterbelts on about a million acres of prairie-plains region farmland in a 100-mile-wide strip, roughly following the ninety-ninth meridian, beginning in the Texas Panhandle and up to the Canadian border. These shelterbelts, drawn on a map as a mammoth 2,000-kilometer-long strip, immediately attracted skepticism and condemnation. Raphael Zon, head of the Forest Service's Lake States Experiment Station who led the project, was aware that the simplistic design was responsible for some of this outcry but insisted that better communication of actual planting patterns would answer to the voices antagonizing large-scale state interference.[88] At the same time, he frankly admitted that the climatic effects of the shelterbelt program over a large territory remain uncertain and that the greatest benefit of shelterbelt planting may be expected when "superimposed on an already existing agricultural economy". In other words, the prospects of continental-scale forest designs were first and foremost social, raising the "still primitive and hazardous existence" of settlers in the Plains region to "a higher level of permanence and stability" (Figure 3.8).[89]

Zon's admittance is probably true to many other regional schemes involving designed forests as key components, regardless of their economic or environmental rhetoric. These schemes most often overlapped with frontier regions in which peoples and species migrated in unforeseeable and therefore uncontrollable patterns. As this chapter shows, there is a solid and long-standing alliance between forest form and regional stability. But in this alliance, the idea of the state as an identifiable and stable community of people had frequently been acting as a silent partner. Since the eighteenth century, Linnean principles scientifically substantiated the claim that distinct species emerge in correlation with specific environments, establishing of the division between endemic and invasive species. Later, locational theories supplied an alibi for creating stable economic circles that clearly define their beneficiaries from those left outside the circle. Shelterbelt programs, proposed in different periods and various geographies, display the common ambition of centralized power forms to make clear distinctions between different peoples.[90] Even more recent large-scale forest plans focused on a limited number of species to prioritize their perseverance over countless others. What it all boils down to is a relentless human desire to separate "us" from "them", to stabilize a territory in which "our" people would be inclined to settle and remain settled for ages.[91] What could be a better symbol of the realization of this lofty goal than a planted forest that would last for a hundred or a thousand years?

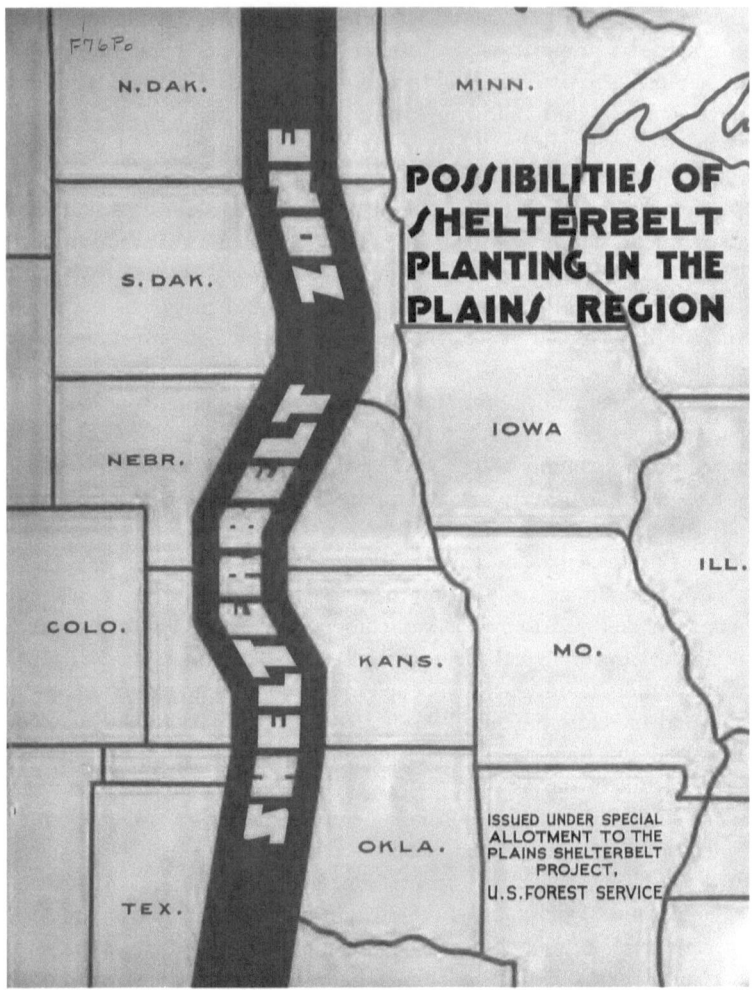

FIGURE 3.8 Diagram of shelterbelt area, from US Forest Service publication *Possibilities of Shelterbelt Planting in the Plains Region*, 1935.

Notes

1 The ceremonial nature of the installation was enhanced by the eventual reuse of part of the dead trees as planks for boats, built by teenagers in the Bronx. Small, Zachary. "Maya Lin's Dismantled 'Ghost Forest' to Be Reborn as Boats." *The New York Times*, November 24, 2021.

2 Fei, Songlin, Johanna M. Desprez, Kevin M. Potter, et al. "Divergence of Species Responses to Climate Change." *Science Advances* 3, no. 5 (2017): e1603055.

3 Dubey, Bhasha, Ram R. Yadav, Jayendra Singh, and Rajesh Chaturvedi. "Upward Shift of Himalayan Pine in Western Himalaya, India." *Current Science* 85, no. 8 (2003): 1135–36.

4 Davis, Margaret B., and Ruth G. Shaw. "Range Shifts and Adaptive Responses to Quaternary Climate Change." *Science* 292 (2001): 673–79, and Fei "Divergence of Species Responses to Climate Change."

5 Ennius, Caecilius. *Remains of Old Latin, Volume I: Ennius. Caecilius.* Translated by E. H. Warmington. Loeb Classical Library 294. Cambridge, MA: Harvard University Press, 1935, 539.

6 Kauṭilya's work, compiled and edited from earlier sources to provide rulers with a definitive guide to statecraft, includes strikingly modern ideas of forest management, including the establishment of tree plantations, forest produce factories, and a hierarchy of forest administrators. On the dating of the *Arthaśāstra*, see Olivelle, Patrick. *King, Governance, and Law in Ancient India.* United Kingdom: Oxford University Press, 2012, 27. On the organization of forests, see the discussion on the roles of forest "superintendents", Olivelle, Patrick. *King, Governance, and Law in Ancient India.* United Kingdom: Oxford University Press, 2012, 140.

7 Staudinger founded the first, short-lived agricultural college in Germany in Groß Flottbek, close to Hamburg, which von Thünen attended between 1802 and 1804. He later supported his former student in working on his book. See Johnson, Hildegard Binder. "A Note on Thünen's Circles." *Annals of the Association of American Geographers* 52, no. 2 (1962): 214.

8 Schumacher, Hermann. *Johann Heinrich von Thünen: ein Forscherleben.* Rostock: G. B. Leopold, 1868, 15.

9 Thünen, Johann Heinrich von, and Peter Hall. *Isolated State; An English Edition of Der Isolierte Staat.* 1st ed. Oxford, New York: Pergamon Press, 1966, 7.

10 The major importance of von Thünen's work to economists is found in his analytic framework, which pioneered a combination of mathematics and marginal analysis to discuss the maximum possible values of products, and in the association of his theory of rent, which largely follows (in the revised edition of the *isolated state*) that of David Ricardo in modifying Smith's position, with his theory of location. That is, von Thünen asserts that rent is primarily a function of distance from the consuming center. See Chisholm, Michael. "Von Thünen Anticipated." *Area* 11, no. 1 (1979): 37–39 and Leigh, Arthur H. "Von Thünen's Theory of Distribution and the Advent of Marginal Analysis." *Journal of Political Economy* 54, no. 6 (December 1946): 481–502.

11 Reception of von Thünen's ideas in economic circles was gradual due to the innovative and insular nature of his work. Yet even a century later, some economists still complained about his "exaggerated" influence in economic geography and planning circles. See Clark, Colin. "Von Thunen's Isolated State." *Oxford Economic Papers* 19, no. 3 (1967): 370.

12 Lösch, August. *The Economics of Location*. New Haven: Yale University Press, 1954, 508.

13 The theory held that "interest should be earned on land, timber capital, and silvicultural expense". See Perry, David A. "The Scientific Basis of Forestry." *Annual Review of Ecology and Systematics* 29 (1998): 437.

14 Von Thünen, *Isolated State*, 112.

15 Von Thünen, *Isolated State*, 107.

16 Cairns, Robert D. "Faustmann's Formulas for Forests." *Natural Resource Modeling* 30, no. 1 (2017): 52.

17 Von Thünen, *Isolated State*, 121.

18 Von Thünen, *Isolated State*, 122.

19 Von Thünen, *Isolated State*, 122.

20 Cairns, "Faustmann's Formulas for Forests," 52.

21 Von Thünen duly recognized this when writing in defense of his initial and most important construct:

> I hope that the reader … will not take exception to the imaginary assumptions I make at the beginning because they do not correspond to conditions in reality, and that he will not reject these assumptions as arbitrary and pointless … this method of analysis has illuminated – and solved – so many problems in my life, and appears to me to be capable of such wide spread application, that I regard it as the most important matter contained in all my work.
>
> *Von Thünen*, Isolated State, *3*

22 Hall was perhaps biased, being the editor of the English version of the *isolated state*. And yet his observation regarding a "extraordinarily completely developed" model bears significance. Von Thünen, *Isolated State*, xxi.

23 The prime example would be James C. Scott's striking argument about the outcomes of such forestry, through which "the German forest became the archetype for imposing on disorderly nature the neatly arranged constructs of science". Scott, James C. *Seeing Like a State: How Certain Schemes to Improve the Human Condition Have Failed*. New Haven: Yale University Press, 1998, 15.

24 Möhring, Bernhard. "The German Struggle Between the 'Bodenreinertragslehre' (Land Rent Theory) and 'Waldreinertragslehre' (Theory of the Highest Revenue) Belongs to the Past – But What Is Left?" *Forest Policy and Economics* 2, no. 2 (2001): 195.

25 Biltmore Forest and its eventual economic failures are discussed in another chapter of the present work. Pinchot, Gifford. *Biltmore Forest: The Property of Mr. George W. Vanderbilt; an Account of Its Treatment, and the Results of the First Year's Work*. Chicago: R.R. Donnelley & Sons Co., 1893, 45.

26 Richardson, Elmo. "'The Compleat Forester': David T. Mason's Early Career and Character." *Journal of Forest History* 27, no. 3 (1983): 115.

27 Parry, B. Thomas, Henry J. Vaux, and Nicholas Dennis. "Changing Conceptions of Sustained-Yield Policy on the National Forests." *Journal of Forestry* 81, no. 3 (March 1983): 150.

28 Mason followed a career trajectory that was uncommon for his time but seems very contemporary today. He started off as a forest ranger but soon turned his attention to writing reports on ownership and production patterns of the "inland empire", which led him to teaching at Montana State University and then the University of California. He then led the newly created Timber Valuation Section of the Bureau of Internal Revenue's Natural Resources Division, which gave him further insights and connections with industry leaders. In 1923, he left the public sector to open a pioneering forestry consulting business in Oregon, out of which most of his later work with both the private and public sectors was conducted. See Maunder, Elwood R., and David T. Mason. "Memoirs of a Forester: An Excerpt from Oral History Interviews with David T. Mason." *Forest History Newsletter* 10, no. 4 (1967): 6–35.

29 Mason, David T. "Comments on the Report of the Committee for the Application of Forestry." *Journal of Forestry* 18, no. 3 (March 1920): 232–5.

30 Mason, David T. "Sustained Yield and American Forest Problems." *Journal of Forestry* 25, no. 6 (October 1927): 625.

31 Mason, David T., and Donald Bruce. *Sustained Yield Forest Management as a Solution to American Forest Conservation Problems.* Portland, Or: Mason & Stevens, 1931, 2.

32 Mason, David T., and Donald Bruce. *Sustained Yield Forest Management as a Solution to American Forest Conservation Problems.* Portland, Or: Mason & Stevens, 1931, 5.

33 Robbins, William G. "The Great Experiment in Industrial Self-Government: The Lumber Industry and the National Recovery Administration." *Journal of Forest History* 25, no. 3 (1981): 130.

34 Mason, David T., and Elwood R. Maunder. "Oral History Excerpts: Memoirs of a Forester: Part II." *Forest History Newsletter* 13, no. 1/2 (1969): 38.

35 Sustained yield was often presented to the public as part of a "New Deal for American Forests". See "The New Deal in Forestry Aims at Sustained Yield Management." *The Science Newsletter* 25, no. 675 (March 17, 1934): 167. For a detailed account of the development of the Lumber Code Authority and Mason's involvement, see Robbins, "The Great Experiment in Industrial Self Government".

36 The cooperative unit type was meant to mingle private and public lands under cooperative management, while the federal unit type was to use only public forests as a source of timber.

37 Public Law 273. "Establishment of sustained-yield units to stabilize forest industries, employment, communities and taxable wealth." 78th Cong., 2d sess., March 29, 1944, ch. 146, §1, 58 Stat. 132.

38 For a piercing critique of the program and the roles the Forest Service played in its crash, see Clary, David A. "What Price Sustained Yield? The Forest Service,

Community Stability, and Timber Monopoly under the 1944 Sustained-Yield Act." *Journal of Forest History* 31, no. 1 (1987): 4–18.

39 Beyond the Shelton cooperative unit, other units were all federal units and included Vallecitos, New Mexico (Carson National Forest), Grays Harbor, Washington (Olympic National Forest), Flagstaff, Arizona (Coconino National Forest), Lakeview, Oregon (Fremont National Forest), and Big Valley California (Modoc National Forest).

40 Hoover, Roy O. "Public Law 273 Comes to Shelton: Implementing the Sustained-Yield Forest Management Act of 1944." *Journal of Forest History* 22, no. 2 (1978): 90.

41 "Hearing Record, Shelton Cooperative Sustained Unit, Hearing, 19 September 1946," Timber Management Office Permanent Files, Pacific Northwest Region.

42 Lyle Watts, Chief of the Forest Service, quoted in Clary, "What Price Sustained Yield?", 8.

43 Stevens, James. "The Simpson Lookout." *American Forests* 60 (February 1954): 19–24, and Collins, Chapin. "Twenty Five Years Later: The Circle that Works [Shelton Working Circle]." *American Forests* 77 (October 1971): 15–19, 58–60.

44 Clary, "What Price Sustained Yield?", 18.

45 Among other things, MacArthur relies in his model on earlier work on stochastic models in mathematics, meant to account for random variables in a system. See MacArthur, Robert. "Fluctuations of Animal Populations and a Measure of Community Stability." *Ecology* 36, no. 3 (1955): 533.

46 MacArthur, Robert. "Fluctuations of Animal Populations and a Measure of Community Stability." *Ecology* 36, no. 3 (1955): 534.

47 "The increase in number of species with area is more rapid in the case of isolated islands or archipelagos than in expanding sample areas on a single land mass." MacArthur, Robert H., and Edward O. Wilson. *The Theory of Island Biogeography*. Princeton, N.J.: Princeton University Press, 1967, 10.

48 The illustration was taken from the work of plant ecologist John Curtis, describing the progression of deforestation and human modification of the landscape around the Cadiz township in Wisconsin between 1831 and 1950. MacArthur, Robert H., and Edward O. Wilson. *The Theory of Island Biogeography*. Princeton, N.J.: Princeton University Press, 1967, 18.

49 For a critical history of the links between checkerboard forests and forest corporations, see Draffan, George, and Derrick Jensen. *Railroads and Clearcuts: Legacy of Congress's 1864 Northern Pacific Railroad Land Grant*. Spokane, WA: Inland Empire Public Lands Council, 1995. For a detailed account of one such grant and the immense legal and managerial complications that followed, see Richardson, Elmo. *BLM's Billion-Dollar Checkerboard: Managing the O and C Lands*. Santa Cruz, Calif.: Washington, D.C.: Forest History Society, 1980.

50 Orians, Gordon. "New Forestry and the Old-Growth Forests of Northwestern North America: A Conversation with Jerry F. Franklin." *Northwest Environmental Journal* 6, no. 2 (1990): 460.

51 The United States entered the IBP program in 1968, a decision that prompted a focus on an ecosystem approach. Throughout its existence it involved 1,800 scientists and tens of millions of dollars in funding, thus promoting the idea of big science and big ecology. See Hagen, Joel. *An Entangled Bank: The Origins of Ecosystem Ecology.* New Brunswick, N.J.: Rutgers University Press, 1992, 164.

52 Handel, Dan. Interview with Jerry Franklin. [Online interview]. February 12, 2024.

53 Handel, Dan. Interview with Jerry Franklin. [Online interview]. February 12, 2024.

54 The irregularity and individualistic nature of old-growth trees are considered inherent and characteristic of the complexity of the system. Franklin, Jerry F., Kermit Jr. Cromack, William Denison, Arthur McKee, Chris Maser, James Sedell, Fred Swanson, and Glen Juday. "Ecological Characteristics of Old-growth Douglas-fir Forests". U.S. Department of Agriculture, Forest Service, Pacific Northwest Forest and Range Experiment Station, 1981, 21.

55 Franklin notes that "the theoretical work that MacArthur was doing didn't have any influence on us whatsoever. We just wanted to understand how a natural forest work". Handel, Interview with Jerry Franklin.

56 Franklin notes that "the theoretical work that MacArthur was doing didn't have any influence on us whatsoever. We just wanted to understand how a natural forest work". Handel, Interview with Jerry Franklin.

57 At the time, the Forest Service was pursuing a pattern that diverged from clear-cutting of large tracts and employed dispersed patch clear cuts 40 hectares in average size, which followed the building of its road systems.

58 The results were published in article form in the inaugural volume of the journal Landscape Ecology and became something of a classic. Forman, Richard, and Jerry Franklin. 1987. "Creating Landscape Patterns by Forest Cutting: Ecological Consequences and Principles." *Landscape Ecology* 1, no. 1 (July 1987): 5–18.

59 Handel, Interview with Jerry Franklin.

60 For an overview and evaluation of the plan, see Thomas, Jack Ward, Jerry F. Franklin, John Gordon, and K. Norman Johnson. "The Northwest Forest Plan: Origins, Components, Implementation Experience, and Suggestions for Change." *Conservation Biology* 20, no. 2 (April 2006): 277–87. For a critical account, see Thomas, Jack Ward. "Sustainability of the Northwest Forest Plan – Dynamic vs. Static Management." Paper presented at the Forest Service Pacific Southwest Region Review of the Northwest Forest Plan Implementation, June 2003.

61 Mapes, Lynda. "Meet the Eminent Scientist, Now 84, Who Vowed as a Boy to Protect Washington's Old-Growth Forests." *The Seattle Times*, July 18, 2021.

62 Agenda 21 was a product of the Rio de Janeiro Earth Summit of 1992, designed as an action plan that could outline practices of sustainable development at the local, regional, and national levels. This was attempted by focusing on both social and economic factors and proposing better resource conservation and management.

63 Rumming, Karin. *Hannover Kronsberg Handbook: Planning and Realisation.* Stand: März 2004. Hannover: Landeshauptstadt Hannover, Umweltdezernat, Baudezernat, 2004, 22.

64 For an overview of some of these forest models, specifically in the context of Swiss forestry, see Handel, Dan. "Forest Hiding." In *Touch Wood: Material, Architecture, Future,* edited by Carla Ferrer, Thomas Hildebrand, and Celina Martinez-Cañavate, 58–61. Zurich: Lars Müller Publishers, 2023.

65 Handel, Dan. Interview with Jacques Herzog. [Online interview]. November 5, 2019.

66 Handel, Dan. Interview with Jacques Herzog. [Online interview]. November 5, 2019.

67 Jury members Gerald Scheff and Cornelia Oberlander commented that it would have been difficult to pick a winner had Tree City was not there, as other proposals did not come close in terms of vision and strategic thinking. See Polo, Marco. "Environment as Process." *Canadian Architect* 45, no. 10 (2000): 16.

68 Bruce Mau argued that the scheme was "not a design at all; it's a recipe or strategy for a series of operations at a meta-level". Polo, Marco. "Environment as Process." *Canadian Architect* 45, no. 10 (2000): 16.

69 Only a fragment of the scheme had been materialized. A new framework plan, submitted in 2021 as part of an official plan amendment process, largely changed the underlying principles of the OMA, Mau, and Oleson Worland proposal. See City of Toronto. "Proposed Redevelopment of the Downsview Lands." Toronto, December 2023. 2nd ed. Accessed February 21, 2024. https://www.id8downsview.ca/_files/ugd/4ea6e4_b688d437659749c48a89d a5947534b37.pdf.

70 The authors note that this is a conservative estimate, as the Ethiopian government has records of 35,000 churches in the highlands. Aerts, Raf, Koen Van Overtveld, Eva November, Alemayehu Wassie, Abrham Abiyu, Sebsebe Demissew, Desalegn D. Daye, et al. "Conservation of the Ethiopian Church Forests: Threats, Opportunities and Implications for Their Management." *Science of the Total Environment* 551 (May 2016): 410.

71 See Sheridan, Michael J. "The Environmental and Social History of African Sacred Groves: A Tanzanian Case Study." *African Studies Review* 52, no. 1 (2009): 74. For a rounded perspective on this issue from diverse disciplinary points of view, see Sheridan, Michael J., and Celia Nyamweru. *African Sacred Groves: Ecological Dynamics & Social Change.* Oxford: Athens, OH: Pretoria: James Currey; Ohio University Press; Unisa Press, 2008.

72 These opposing styles of Christian teaching, one anchored in Orthodox Church hierarchy and the other based on charismatic monks, remained a source of conflict for over a millennium. See Crummey, Donald. "Church and Nation: The Ethiopian Orthodox Täwahedo Church (From the Thirteenth to the Twentieth Century)." In *The Cambridge History of Christianity*, edited by Michael Angold. Cambridge History of Christianity. Cambridge: Cambridge University Press, 2006, 461.

73 Hable Selassie, Sergaw. *The Church of Ethiopia: A Panorama of History and Spiritual Life*. Addis Ababa: A Publication of the Ethiopian Orthodox Church, 1970.

74 The precise chronology of forests in the Ethiopian highlands is not easy to ascertain. Ecologist Meg Lowman notes that

> there is not the luxury of historical data to answer this ... there is wonderful anecdotal memory from some of the elder priests ... [and] we do have soil cores that indicate that approximately 50% of the northern half of the country was blanketed in forests, before humans started clearing them.
>
> *Handel, Dan. Interview with Meg Lowman. [Email interview].*
> *February 25, 2024*

75 Eshete, Alemayu Wassie. "Ethiopian Church Forests: Opportunities and Challenges for Restoration." PhD diss., Wageningen University, Wageningen, The Netherlands, 2007.

76 Wassie identified 168 tree species and marked those that fell under international definitions of endangered species.

77 Wassie's thesis adviser, tropical forest ecologist Frans Bongers, possibly recommended his paper, as he was involved with and was later president of the ATBC, in which Lowman was a treasurer.

78 Lowman, Meg. *The Arbornaut: A Life Discovering the Eighth Continent in the Trees Above Us*. New York: Farrar, Straus and Giroux, 2021, ch. 11, Kindle.

79 Lowman, Meg. *The Arbornaut: A Life Discovering the Eighth Continent in the Trees Above Us*. New York: Farrar, Straus and Giroux, 2021, ch. 11, Kindle.

80 Lowman, Meg. *The Arbornaut: A Life Discovering the Eighth Continent in the Trees Above Us*. New York: Farrar, Straus and Giroux, 2021, ch. 11, Kindle.

81 The 2016 study found that the round church forests were more diverse and less degraded, containing more indigenous conifers than the forests with a more complex shape. See Aerts, Raf, Koen Van Overtveld, Eva November, Alemayehu Wassie, Abrham Abiyu, Sebsebe Demissew, Desalegn D. Daye, et al. "Conservation of the Ethiopian Church Forests: Threats, Opportunities and Implications for Their Management." *Science of the Total Environment* 551 (May 2016): 404.

82 Most of the investment in the GGW comes from the World Bank and the multilateral Global Environmental Facility fund. See Climatekos GmbH. "The Great Green Wall: Implementation Status and Way Ahead to 2030." Published online by United Nations Convention to Combat Desertification (UNCCD), November 2020, 28.

83 Influential CCP leader Gao Gang initiated the shelterbelt in northeast China as part of its transformation into a center of heavy industry. Solecki, Jan. "Forest Resources and Their Utilization in Communist China." *The Forestry Chronicle* 40, no. 2 (June 1964): 235.

84 Solecki, Jan. "Forest Resources and Their Utilization in Communist China." *The Forestry Chronicle* 40, no. 2 (June 1964): 237.

85 The propaganda blitz that accompanied the Great Plan included a commissioned oratorio by Dmitri Shostakovich, which hailed the forestation of the steppes and Stalin's role in the transformation of nature.

86 In the late nineteenth century, the Russian steppes were portrayed by Russian intellectuals and administrators as "a land of fertile soil, opportunity, prosperity, freedom, beauty and Russianness." See Moon, David. "The Environmental History of the Russian Steppes: Vasilii Dokuchaev and the Harvest Failure of 1891." *Transactions of the Royal Historical Society* 15 (2005): 156.

87 See Bates, Carlos G. "Windbreaks: Their Influence and Value." *Department of Agriculture, Forest Service – Bulletin 86*, September 30, 1911, 12.

88 Zon, Raphael. "Shelterbelts – Futile Dream or Workable Plan." *Science* 81, no. 2104 (1935): 391.

89 Zon, Raphael. "Shelterbelts – Futile Dream or Workable Plan." *Science* 81, no. 2104 (1935): 394.

90 The correlations between ethnic conflicts and forest shelterbelt planting are striking. In the Western steppe, The Russian Empire was attempting to establish the presence of Slavic over Asiatic people as it conceived the first protective forests. Throughout different tree planting schemes, Northeast China was a dynamic conflict area between Manchu, Chinese, and Mongols and was later wrestling Japanese and Russian influences. In current-day Ethiopia, the Green Wall coincides with the destabilized Tigray region.

91 This underlying desire can perhaps account for the curious homology between advanced environmental and conservative political agendas, focused on resisting uncontrolled "invasive" migration of human and nonhuman species. Journalist Sonia Shah noted that under this logic "migration is by necessity a catastrophe, because it violates the natural order". See Shah, Sonia. *The Next Great Migration: The Beauty and Terror of Life on the Move.* New York: Bloomsbury Publishing, 2020, 62.

4

ECOLOGICAL HAVENS

At the end of the first decade of the twenty-first century, architecture seemed to be at a loss. Aligning itself with market globalization for most of the previous decade and sharing the delights of private jets and accommodating contracts in authoritarian countries, it suddenly found its salesmanship hindered by the carelessness of actual people, failing to return their loans. Not only did the source of the subprime mortgage crisis lay in proximity to the most basic architectural product, the house, which made it harder to sweep under a rug of deliberate amnesia, but the shockwaves hitting major banks across the world soon also made it clear that the stream of easy money that was sponsoring slender condominium towers in Toronto, opulent casinos in Macau, and instant museums in Shanghai was coming to an abrupt end. Losing its supply of lucrative commissions, architecture was experiencing a variety of withdrawal syndromes. As billings plummeted, firms in the global North went on rounds of layoffs and budget cuts, and academics gathered in soul-searching support groups, atoning for the mistakes of others, and bravely condemning the rapacious order of plutocrats who had done everything wrong, apart perhaps from an incidental benevolent gift to a research institution genuinely advancing human knowledge. Some of the influential emissaries of architecture vowed to change their ways, while others took on the task of finding a new mission for it. If buildings alone could not be trusted, perhaps it was time to venture into natural systems, integrate them into spatial design, and so redefine its goals and meaning. Thus, the idea of Ecological Urbanism was presented to the world. Its official birth certificate carried the mark of Harvard, and the announcement took place at the Graduate School of Design in the spring of 2009, with a major conference and a publication overseen by Dean Mohsen Mostafavi.

In describing what Ecological Urbanism was, the dean weaved an argument based on the philosophical ideas of French thinkers Félix Guattari and Jacques Rancière. Building on their concepts, he made the case that

DOI: 10.4324/9781003473411-5

the fusing of ecology and urbanism would yield endless opportunities for spatial design to do what it likes to do: engage in speculation. "We need to view the fragility of the planet and its resources", he wrote, "as an opportunity for speculative design".[1] Such "creative imagining", as was implied by projects included in the publication and conference, would encompass everything that is right and true: be flexible and adaptive, address the social disparities in the favelas of Rio, alter the course of careless developments in Gulf cities, reuse the broken infrastructure of New Orleans, learn from the markets of Lagos, enhance democratic principles, and get industrialized societies back on track to a timely compliance with the Kyoto Protocol.

The intention to mint of a new paradigm was evident in the list of names invited to the conference, clearly curated to bestow the aura of a memorable event. The plenary speaker was Rem Koolhaas, the high priest of the intellectual side of the profession, brought in to preach as part of his duties as professor in absentia. As Koolhaas, a bald, towering figure, made his way through the mass of believers in the dim entrance hall, there was a feeling that architecture may be saved after all. His sermon turned out to be less hopeful. The wit was there when he apathetically commented, "[W]ell, this is over", as a big black X appeared on a collage of shiny glass towers representing his firm's flirtations with Gulf states and Asian clients. But his self-consciousness did not yield anything this time. An epoch was coming to an end, but the oracle from Rotterdam clearly had no idea what was coming with the tide. This unhappy beginning marked the opening of three days of gathering, in which presenters did their best to legitimize the coming of Ecological Urbanism. However, these efforts, often adorned with the best talent and eloquence, only contributed to a growing sense of confusion among attendees. Certain speakers dragged sustainability into the ring, while others put their trust in technology. Exotic ecologists were brought in from their distant lands to speak about complexity and biotic dynamics, but the moderators kept turning these hard-earned insights into obtuse metaphors of city management.[2]

Ecological Urbanism as a coherent program was largely abandoned by architectural thinkers and practitioners since then. However, one project presented on the occasion proved to have a sustained impact. This was a draft proposal to flood Milan with forest vegetation to the point of transformation, presented by Stefano Boeri during the second day of the conference. The main point made an appearance around slide fifteen, which featured a press photo of a fox jumping over a coffee table in a European city. This moving photo was there to illustrate a new relationship between nature and cities, and Boeri was speaking about the need to recognize the autonomy of nature and the limits of human control.

But what was pulsating under his argument felt very familiar: it was a rehearsal of the ages-old idea of the non-domesticated world coming back to conquer civilization. Boeri's proposition was designed to counter this process: if natural systems were swirling out of control, he was planning to be there at the gates of Milan with his chimera of green design and market development that would contain nature in vertical forest reserves. A decade later, *Bosco Verticale* is branded as the magic leap into the era of environmental awareness and as a model that, after being already built in Milan, can be exported to be planted in such diverse climates, cultures, and political systems as Nanjing, Eindhoven, Chicago, and Cairo.[3] Boeri calls this model "metropolitan reforestation", betraying his belief in a sylvan past that unites the civilizations of the world, including the ones growing out of the sands of Arabia (Figure 4.1).[4]

FIGURE 4.1   Scheme for Biomilano, Stefano Boeri Architetti, circa 2010. © Stefano Boeri.

But the actual Bosco Verticale, as opposed to the manifesto, is something entirely different. Under the veils of eco-talk and Romantic idealism lies a shrewd real-estate scheme. The cadre of investors in the project brought together representatives of the thousand faces of global money, including private equity company Hines, financial services provider TIAA, holding company Domo Media, and insurance company Milano Assicurazioni, most of which demonstrated a cool indifference toward environmental issues in previous investments. The lure of the project for the financiers was not in the forest but in what was underneath the trees, two high-profile residential buildings for the upper class that came complete with the marketing benefits of a new Tesla: technological, elitist, and comfortably environmental.[5] In that, Bosco Verticale ingeniously crystalizes contemporary metropolitan consciousness. It is home to people who, in most cases, eagerly consume products that take their environmental toll on the other side of the earth while allowing themselves to dwell in a simulated forest with questionable environmental benefits, designed for the immaculate display of conscious and responsible living.

Both the investors and the astute technical experts involved with the design and construction of Bosco Verticale had put their faith in numbers. The project boasts 800 trees, 5,000 shrubs, and 11,000 floral plants integrated in its cantilevered concrete balconies. Resourceful number crushers calculated that the project's plant life is equal to 10,000 square meters of forestland.[6] However, seeing a big oak lifted twenty-eight floors up in the air should make one question this claim. By any comparison – species diversity, resilience in a major disturbance scenario, or ecosystem services – the project would be on the losing side when put against even a minimally managed forest tract. While assembling a large number of trees close together allows for convincing carbon emission offsetting calculations, it hardly constitutes a sound ecological argument. In this regard, proponents of eco-verticality often confuse forest with landscape painting, neglecting to realize that the environmental usefulness of forest environments stems as much from the soil as from visible greenery (Figure 4.2).

This blind spot is perhaps not surprising when considering the ancestry from which Boeri's project emerged. Rather than ecological or biological sciences, it has its origins in internal architectural and urban conversations about the possibility of creating "green architecture". While environmental concerns, and specifically energy consumption, were recurring issues in architectural circles since at least the 1960s, the emergence of green architecture cannot be attributed to a single lineage. In their attempt to critically recount its history, A. Senem Deviren and Phillip Tabb proposed to understand the many tributaries that contributed and shaped such architectures as part of an ongoing process of

FIGURE 4.2   Lifting a tree to the upper floor during construction of Bosco
Verticale, 2012. © Marco Garofalo.

"greening", resulting from the attempt to transform modern architecture.
"The greening of architecture", they write, "is a mutable and iterative
process that occurs, changes, builds intelligence, and evolves over time."[7]
This evolution means that environmentally conscious terms such as *green
architecture* or *sustainable design* were often used loosely and meant different things in different times. As both mainstream architectural media and
architectural practice caught up with the term in the late 1980s, green
architecture spread globally as it gradually lost any specific meanings.
In his text for the Ecological Urbanism publication, architect Julien de
Smedt commented that "'green' and 'sustainability' … have become dan-
gerously afloat in ambiguity and indeterminacy. Sustainable architecture
is everywhere and nowhere".[8]

However, the type of architecture that inspired Bosco Verticale can be
traced to a specific subset of green architecture, which was defined during the late 1970s in "developed" countries, following a series of energy
crises and deepening skepticism toward the capacity of the planet to sustain life as before. These anxieties found expression in the works of individuals who reckoned what they saw as architecture's broken relationship
with nature. Looking back in 2000, architect James Wines, attempting to
separate the wheat of worthy green architecture from the chaff of technical sustainable problem-solving, warned against the normal experience of
visiting a

well-publicized "ecological building" and be handed a checklist brochure of its earth-friendly virtues, while there is no visible evidence
of any attempt on the part of the designer to resolve these contributions in terms of art. It may be green, but it is boring architecture.[9]

The type of architecture he was hoping for would have to shed the stylistic
baggage of misguided modernism, focused as it was on industrial metaphors, and develop a new "ecologically-based architectural language".[10]
    Wines had been grappling with these issues for some time. In 1980,
his firm SITE proposed an unlikely project for a Best Products Company
department store in Richmond, Virginia. Its idea was to maintain an
existing stand of oaks and undergrowth and let it enter the building by
designing a space that fractured the otherwise simple box structure.[11]
The front of the building was separated from the rest by this new space of
greenery through which shoppers, crossing a bridge, entered the show-
room. Wines, who started his career as a sculptor, had set up shop in New
York City's SoHo of the late 1960s as a cadre of anti-art artists were testing
out their ideas on the canvas of the neighborhood's abandoned industrial
debris.[12] In many ways, the BEST "forest building" was indebted to these
ideas in its anti-building attitude and in the explicit homage to Gordon
Matta-Clark's Building Cuts series. But no less significant was the project's conscious divergence from the protocols of modern architecture,
which would habitually require the wiping out of everything on-site as
a means of achieving a preliminary clean slate. This position was amplified in renderings that show the building completely surrounded by tree
palisades with even its most important commercial signifier, the BEST
logo, hardly visible.[13] These depictions resist functionalism, clear form,
the distinction between building and nonbuilding, and the entire stylistic
patrimony of modern architecture. There was, however, something else
in the drawings. SITE's deliberate attempt to create the appearance of
a building "invaded and consumed by nature" inform a certain anach-
ronism which makes it hard to ascertain not where but when exactly
is the project situated.[14] Is it a glimpse into the primordial forests that
once covered Virginia or into a future in which the edifices of consumer
society would be overtaken by plants? In retrospect, Wines would argue
that this ambiguity was designed to evoke "nature's revenge".[15] But even
if this statement is imbued with later discussions about architecture and
environment, it still throws in the temporal aspect, suggesting that the
meeting points between them are defined cyclically by forests that are
taken down only to reappear in a later point in time, and turn building
into ruin. The forest building was in real time a bold comment on both

architectural and consumption cultures. Its only problem was it played too deep in the waters of postmodernist sign language to be taken to heart. And so it was relegated to the curiosity section next to other SITE designs for BEST, which included an asphalt parking surface becoming the roof of a building, and an enclosed tropical environment as part of the shopping experience (Figure 4.3).[16]

Then came Emilio Ambasz, emerging as a chief herald and connoisseur of green architecture. In the 1960s and early 1970s, the Argentinian architect quickly amassed cultural capital through his influential tenureship as curator of design at the Museum of Modern Art (MoMA). He then turned his attention to imagining buildings that would be covered in green. This focus had less to do with a newly found planetary conscience and more with an intellectual duel he was leading with the academic and professional elites of North American postmodern architecture.[17] During the 1970s, several figures in these circles were producing a body of work heavily influenced by semiotics (in the case of Peter Eisenman), and French structuralism (Diana Agrest and Mario Gandelsonas), putting forth spatial concepts that portrayed architecture as an autonomous and hermetic realm.[18] Another group, centered on architectural historian Vincent Scully and influenced by Denise Scott Brown and Robert Venturi, were

FIGURE 4.3   Model of Forest Building for Best Products company, James Wines/SITE, 1980. © SITE.

125

channeling their interest in architectural sign systems into projects that were obsessed by classical orders and historical ornaments.[19] Ambasz, on his part, claimed that they were giving the wrong answers. "My way of doing ornament" he would later reflect, "was by using nature".[20] With this, Ambasz began to pave his way to proclaiming himself the forefather of green architecture.

It started with a series of paper projects formatted for exhibitions and the architectural press. His 1982 project for the Schlumberger Research Center in Austin was disguised as a park, in which, as the MoMA press release later put it: "architectural pavilions emerge here and there, like the 'follies' of a traditional garden, to lend an accent".[21] Aptly put, since Ambasz's research center and later green projects, garnished here and there with references to the English or French garden traditions, were designed to be consumed by the higher echelons of society, delighting in their good taste as their carbon-intensive lifestyles continued unabated. Ambasz went on to construct several projects. In the Mycal Cultural and Athletic Center in Shin-Sanda, Japan, built in 1990, he attempted to sink the gigantic structure in a green forested topography. In the 1995 ACROS center in Fukouka, he went further to create an enlarged mountain version of SITE's forest building on a 2-acre site. The fourteen-story building was covered by a step garden made of thirteen tiers, each 6 meters deep. The steps were planted with 37,000 plants of seventy-six species, including trees such as maple and Japanese plum, which drowned the concrete and glass building in billowy green. The project gradually became a hit with both high architectural and sustainable design circles, and later studies explored its heat-reducing capacity and environmental performance.[22] However, its point of departure remained merely aesthetic, as Ambasz's popular dictum "green over grey" well captured (Figure 4.4).[23]

While very different in temperament and approach, both Wines and Ambasz remained trapped in disciplinary parlor games, sketching out a type of green architecture that was largely based on its own preoccupations. While later buttressed by environmental calculations and rhetoric, their forest buildings did not engage with the resulting forest ecologies. Nowhere is this clearer than in their treatment of the ground as a surface covered in trees and undergrowth rather than as a mass. In that, these projects carried forward a long-standing tradition in Western architecture, which was defined for hundreds of years through the abstraction of sites and on what can be regarded as a deliberately limited understanding of the land.[24] The manufactured forests that fill the renderings of such green architecture err by replicating the forest floor, which is an abstraction that dissects the complex life of plants with an architectural scalpel.

FIGURE 4.4   ACROS Center in Fukouka, Emilio Ambasz, 1995. Photograph by Hiromi Watanabe. Courtesy of Emilio Ambasz.

Trees have root systems, and placing them in open caskets, as is done in Bosco Verticale, yields no forest life. But perhaps that is precisely the point. These projects trade not in actual ecologies but in metaphors of a magical forest ecology that can transform cities for the best. Such metaphors are cultivated to be funneled through established financial and developmental structures and applied to highly developed metropolitan contexts in a seemingly progressive yet entirely undisruptive manner.

★★★

There was, however, another meandering path by which the reconceptualization of cities in ecological terms infiltrated building design. In 1995, Malaysian architect Ken Yeang published the book *Designing with Nature*. This was an adaptation of his PhD dissertation, submitted to the University of Cambridge in 1980.[25] The book's title, making a nod to Ian McHarg's famous work, *Design with Nature*, represented Yeang's ambition to expand McHarg's lessons to landscape architects into the realms of architecture and urban design. While in Cambridge, and later at the University of Pennsylvania, Yeang attended courses in ecology, allowing him to construct a solid case for his later theories. His initial argument is one of translation: if ecosystem concepts could be explained in spatial design terms, then extensive environmental damage resulting from built projects would be averted. After articulating several key ideas from ecology and environmental biology, Yeang outlines an ecological approach to design, based on principles of ecosystem analysis and management.[26] Most of the book is then devoted to the description of the "built environment" and its complexities in ecological terms. In the final chapter, Yeang proposes that architectural theory should open its doors to environmental sciences and follow the thread he is sketching, leading to a proper ecological design. This he defines as the outcome of "a design process in which the designer comprehensively minimizes the anticipated adverse effects that the product of that design process has upon the earth's ecosystems and resources".[27] Yeang's approach could have remained an interesting footnote in architectural theory had he not established himself as a pioneering architect in his home country. In 1976, Yeang opened a new architectural partnership with Tengku Robert Hamzah, and during the following decade, he began to apply ecological ideas in his designs. In the late 1980s, he developed a number of schemes for green high-rise buildings, which attempted to integrate an array of bioclimatic technologies and intensive vegetation throughout their different floors. These included the IBM Plaza in Kuala Lumpur, completed in 1987, introducing the idea of a continuous landscape, linking a green area on the ground floor with landscaped areas on the face of the building. This idea was repeated in the well-known Mesiniaga Tower in Subang Jaya, Malaysia, completed in 1992. In Menara Bousted from 1987, balconies were intensively planted to filter and cool the air. During that time, Yeang refined his general theory to become a regionally specific typology which he termed the tropical skyscraper: "the climatically responsive tall building", he writes in 1992, "would enhance its users' aesthetic well-being while enabling them to be aware of and to experience the external climate of the place."[28] The visual component of the work was again emphasized in relation to the function of plants: "the use of tropical planting enhances the aesthetics of the tall building as a tropical structure".[29] Yeang goes on to emphasize the ecological feasibility of intensive planting and its potential contribution to species diversity and healthy ecosystems, but it is clear at this point that purely architectural considerations compete and sometimes override ecological performance.[30] While his sound ecological background never led him to argue that his vertical landscaping and intensive use of trees qualify as forests, his projects' image and interpretations contributed to later designed "vertical forests" that presumed to transform urban ecologies (Figure 4.5).

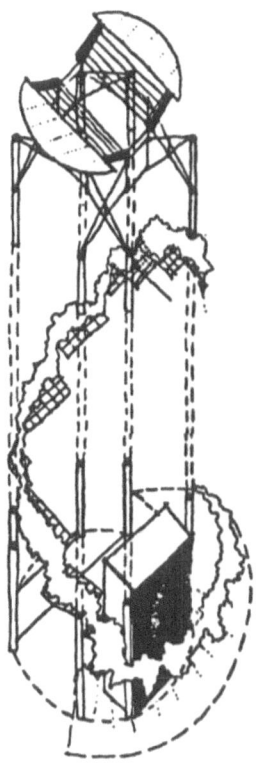

FIGURE 4.5   Planting scheme for Menara Mesiniaga in Kuala Lumpur, Ken
Yeang, 1992. © Ken Yeang (2024).

★★★

If architects remained somewhat confined to their disciplinary misconceptions, perhaps forests could transform cities through the stewardship of
other disciplines under the heading of the urban forest. The contradictory
term was coined in 1965 by Danish-Canadian forester Erik Jorgensen,
who envisioned a "specialized branch of forestry … [that] has as its
objectives the cultivation and management of trees for their present and
potential contribution to the physiological, sociological and economic
wellbeing of urban society".[31] According to Jorgensen, media theorist
Marshall McLuhan considered the term appropriate as "it gave an image
of tying the city, the cultivated and civilized, to the rural and unmanaged

rural areas".[32] In other words, it was a catchphrase that had little to do with the ecologies of either cities or forests. Nevertheless, McLuhan apparently knew what he was talking about, since the term caught up and became popular in various places around the world.[33] The 1970s and 1980s brought exponential growth of activity and public debates around the issue, prompted by an apparent loss of tree cover in American cities. Organizations such as the National Arbor Day Foundation organized massive planting campaigns and promoted Tree City standards, which included "a city forester or tree board responsible for city trees; a tree-care ordinance on the books; an active urban forestry program with a budget equal to two dollars per citizen".[34] The US Forest Service backed the goal of greening the cities by establishing a consortium for urban forestry, which involved its scientists and researchers from leading universities. In 1978, the US government officially referred to the cause of urban forestry in the Cooperative Forestry Assistance Act, designed to avail funding and technical assistance for forestry initiatives on private lands. The act drew a connection between the declining health of forests in cities, the assumed ecological benefits of trees in urban environments, and the potential enhancement of economic value in commercial and residential properties.[35] It also made the link between urban forestry and larger environmental concerns, arguing that urban trees "can aid in reducing carbon dioxide emissions … and reducing energy consumption, thus contributing to efforts to reduce global warming trends".[36] And so, urban forestry was introduced as a magic win–win formula for everyone involved – from mayors to scientists, and from environmentalists to citizens. No less important, it extended the authority of the US Forest Service from scientifically managed federal forests, where it had resided since the early twentieth century, to the complex, multistakeholder realm of urban life.

This change of boundaries alarmed several landscape architects, who saw in the urban forest phenomena a reincarnation of earlier conversations about city trees and their roles in beatification schemes that had been proposed for American cities since the late nineteenth century. This concern awakened a disciplinary feud between foresters and landscape architects. If there were such a thing as a city forest, who would be best equipped to be in charge of its design: an expert with a close familiarity with the structural and botanical characteristics of trees, or one who can focus on their functions and meanings to communities? History rhymes throughout the pages of Landscape Architecture magazine. In 1915, landscape architect Harris Reynolds, referring to the newly emerging function of city forester, wrote: "[W]e would not condemn the forester who holds such a position creditably: but the question is, would he not have done better had he had the training of a landscape architect?"[37] In 1980, landscape architect Russell Beatty, vented his frustrations in a letter to the editors, wrote that

> while working on several articles on the subject of "urban forestry", I was struck by the lack of visibility of landscape architects in this emerging discipline ... the result is that design is given lip service and it often superseded by the scientific factors of urban forestry.[38]

Beatty was not wrong. Circa 1980, he could still make the case that the practice of urban forestry should be systematized through urban tree plans and that landscape architects should be the ones leading the way.[39] But in following years, his discipline was quickly losing ground to scientists who were hard at work to substantiate the quantitative claims of urban forestry. Even though forest ecology had developed significantly by that time, at the end of the 1970s would be urban foresters were not able to answer even the most basic questions about the benefits of trees in highly urbanized areas.[40] They didn't know how many trees were out there in different metropolitan areas, what was their species composition, how did they interact with ecosystems and perhaps most important to decision-makers, how much did an urban tree cost and how much was it worth?[41] Rowan Rowntree, a researcher at the Northeastern Forest Experiment Station who previously attempted to assess the air- and water-cleaning potentials of trees in Dayton, Ohio, began speaking publicly and publishing articles along this vain. "The relationship between people and trees, in our work", he said, "would have to be described as a pecuniary one."[42] The cause was promoted through a series of studies that laid the empirical foundations of urban forests, dealing among other things with the question whether they even qualify as forests. In 1984, Rowntree attempted to establish this point, writing that in temperate regions, it is likely 60–80% of cities support enough trees to meet conventional definitions of "forest".[43] He then moved on to discuss the relationship between land use patterns and the distribution of canopy cover and the much-needed research into issues of species composition and typology of urban forest structures. This was not a smooth transition. Other scientific analogies, such as the urban canopy layer, curiously composed of both trees and buildings, would have to be made between forests and cities to fortify the metaphor and promote the cause of ecologically beneficial urban forests.[44] This cause was peddled in professional conferences and publications and brought to public attention through associations and coalitions so that by 1989, an *LA Times* article could describe a tree masterplan for the city of Thousand Oaks,

California, as "thrusting [the city] into the forefront of a resurgent urban forestry movement".[45]

At the time of writing, this movement was still largely restricted to North America, with European experts, influenced by American examples, gradually joining the conversation and proposing urban forestry projects in their countries.[46] Circumscribed by Rowntree's focus on cities in temperate regions, the spread of urban forestry beyond the West progressed along the lines of climatic regions. In 1992, Guofang Shen, president of Beijing Forestry University, organized the first symposium on urban forestry in China. Impressed by French urban forest cases he saw a year earlier at the World Forestry conference in Paris, Shen sought to apply similar concepts to Chinese cities. Here, too, disciplinary feuds had to be won before urban forestry ideas began to be implemented during the 1990s.[47] In 2004, the national Forest City program was announced, creating a unified index of urban forestry and encouraging planning acts that would "let forests enter into the city and let the city embrace forests". In 1997, the National Environment Council in Chile announced a policy goal of using trees and green areas to fight severe air pollution in the Gran Santiago metropolitan region.[48] Sub-Saharan Africa joined the movement only in the 2000s through the mediation of international organizations. In 2002, the city of Tshwane in South Africa set to plant 115,200 street trees in areas of disadvantaged communities. Johannesburg followed suit when planting 90,000 trees and announced 200,000 more in 2009 as part of the World Cup program.[49] Massive planting campaigns in urban areas then began to appear deeper in the tropical areas of the continent and are currently in different phases of implementation in more than fifty cities.[50]

However, research remains stubbornly inconsistent regarding the actual effects of urban forestry. In the 1980s, scientists looking into nitrogen-cycling processes observed significantly slower rates of nitrification and N mineralization (the process by which organic nitrogen [N] is converted into inorganic forms that plants can absorb) in soil samples taken from urban forests.[51] Such studies were alluding to the fact that the activity of soil microbes and invertebrates was much hindered by the biological heritage of urban land, shaped by decades or centuries of contamination. That, in fact, from an ecological standpoint the urban forest is an entirely different entity from the ones after which it was modeled. The fact that the urban forestry discourse is dominated by North American and European authors and that even after six decades its precise definition remains nebulous, makes it prone to ambiguities and misconceptions.[52] Yet this fact doesn't stop authors from claiming a broad consensus exists on its multiple benefits, thus propelling its global spread.[53] As urban forest schemes appear in regions hitherto unfamiliar with the concept, their reasoning is laden with intangible aesthetic, social, and economic benefits. In its applications in certain African cities, it is argued that the urban forest plays an important spiritual role in the life of communities.[54] And so the urban forest reverts to the status of a sacred grove, and forest ecology simply becomes a useful metaphor for scientists, landscape designers, city administrators, and urban communities seeking to promote their version of the best possible urban environment.

★★★

To realize the ecological urban forest dwells more in the realm of metaphor than in that of scientific fact is also to open up its potential to connect with past and future visions. In 1965, as *urban forestry* was first proposed as a useful term, artist Alan Sonfist began work on Time Landscape – a precolonial forest lot at the corner of La Guardia Place and West Houston Street. Sonfist, who was twenty years old at the time, proposed the project as a homage to an isolated hemlock forest that once existed in the Bronx River and served as his sanctuary before being destroyed by the city.[55] His initial idea was to create fifty monuments that would chart "the life and death of natural phenomena", arguing that "public art can be a reminder that the city was once a forest or a marsh".[56] The medium was to be an assemblage of biological material that could commemorate the way Manhattan Island may have appeared before the Dutch upon their arrival in 1624.

The proposal, sent accompanied by a set of drawings to Mayor Robert Wagner and Comptroller Abraham Beame, was imagining the lifting of concrete and the revealing of nature that had been there before colonization. Beame's office connected Sonfist with Tom Hoving, who was the mayor's park commissioner and would soon begin his tenure as director of the Metropolitan Museum of Art. Hoving was enthusiastic about the idea but urged Sonfist to develop a sounder scientific base, which sent him into more than a decade of development.[57] During that time, he was invited by György Kepes to the Massachusetts Institute of Technoology to develop his idea in the company of researchers and scientists. When Hoving assumed his role at the Metropolitan Museum of Art, he informed Sonfist that the first Time Landscape would be created for the museum, holding that "the future of the American wing lay with the history of this land itself".[58] A decade later, as Hoving left the museum, it was clear that the project would have to happen elsewhere, and through many conversations with Ruth Wittenberg and Jane Jacobs, and other "movers and shakers of the city", Sonfist honed in on Greenwich Village as the site for his forest.[59] One of the supporters of the project who lived nearby was Congressman Ed Koch, and when he became mayor in 1977, the bureaucratic obstacles disappeared, and the forest came into existence.

The realization of the project finally took place on a 45- by 200-foot rectangle, owned by the Department of Transportation, on which Sonfist planted native tree species: gray birch, red cedar, black cherry, witch hazel, oaks, white ash, and elms. However, from the outset, it was made clear that the work was not meant as a manicured park: in the attempt to rep-licate the biotic diversity and aesthetic complexity of the oak–hickory forests that once populated the region, other tree species, shrubs, and ground-covering plants were thrown in the mix. And so pokeweed, milk-weed, tulip trees, bindweed, and catbrier vines made cameo appearances in this piece of ecological theater, in which the artist was determined to make the point as clearly as possible. Organizing the cast of flora in three acts, the lot was parceled from south to north to follow the stages of the forest from grassland to saplings to fully aged trees.

This spatial and temporal organization brings to mind the demonstra-tive qualities of large-scale dioramas of the kind found in Harvard Forest or the Museum of Natural History, a point that did not escape the lukewarm review Time Landscape received in *Artforum* soon after it opened.[60] At the time the review was written, the forest seemed to be growing according to plan. But of course, the living diorama could not be sustained as such. In the years following its planting, not only did its categorical organiza-tion decline (grassland changes, saplings grow into trees), but invaders also made their way in through the steel bars. Author Michael Pollan, writ-ing his first book about a decade after Time Landscape was put in place and walking next to it most mornings on his way to work, confessed his dislike to "this sort of 'garden', of which Emerson and Thoreau would have approved – for the very reason that it's *not* a garden". But then on another occasion, he takes notice of a vine along the fence. It is night-shade, a plant that found its way to America with the arrival of the white man. Feeling victorious, he writes: "Aha! This smug little wilderness is a garden after all."[61] Unless someone with a distinctive eye tends to it, Time Landscape may degenerate into a weed lot. Sonfist argued on several occasions that he is not bothered by the changes brought about by the inherent defects in his plan. Thankfully, the gardeners at NYC Parks and Recreation Department do not adhere to these statements and are taking regular measures to weed and incinerate intruders.[62]

Compromised by weeds or not, Sonfist's intervention took a different path from contemporary artworks with environmental undertones. In its ingrained change factor, it countered the final form that was the goal of the masculine land art of the 1960s, which imprinted its geometries, blew up rocks, or manipulated entire landscapes to create everlasting statements. Even by comparison to more ecologically minded works that tacitly transformed urban landscapes such as Bonnie Ora Sherk's Portable Parks in San Francisco (1970) or Wheatfield – A Confrontation by Agnes Denes in Battery Park (1982), which sought to position the artist as an agent of awareness and potential change, Time Landscape remains obscure and ambivalent. The artist was significantly not present, which led the *Artforum* reviewer to snobbishly comment that he was missing the point of his own work.

Be that as it may, as a piece of reconstructed forest, Time Landscape lucidly demonstrated the aims and methods of its originator. The active transformation of the site into a supposed precolonial sylva aligned it with the burgeoning interest in designing ecosystems that would coalesce in the 1980s into the subdiscipline of restoration ecology. Laura J. Martin, following the many paths taken by American scientists in search of the reintroduction of wildness to ecosystems, posits that restorationists have offered a third way between preservation and conservation, "challenging the idea that a place is either untamed or managed, wild or designed".[63] Her narrative charts the consolidation of these ideas from the early twentieth-century botanical interest in "wild" species to the work of female scientists introducing the restoration of native plants as a main subject in ecological labs to Aldo Leopold's involvement in the University of Wisconsin arboretum. At the dedication of the arboretum in 1934, Leopold stressed that the idea was not to collect plants but rather "to reconstruct ... a sample of original Wisconsin – a sample of what Dane County looked like when our ancestors arrived here during the 1840s".[64] The adherence to historical fidelity would come to characterize much of the rhetoric around restoration ecology and is also apparent in the statements surrounding Time Landscape. But Martin's narrative complicates this by showing that restorative ecology did emerge not only from Leopold's celebrated Land Ethic but also from a fascination with radical redesign schemes, pursued through close communication between scientists and military strategists during the Cold War era. These collaborations sought to prepare for Doomsday by learning from past ecological disasters and envisioning possible paths towards environmental recovery.[65] And so the theory and practice of restoration ecology were established in view of imagined pasts and imagined futures. The ambition to speculatively codesign nature, and to do so by liaising with nonhuman "collaborators", exposed ecologists to doubts and criticisms. Historians wondered why the prairies in the 1840s or the East Coast in 1620 should be considered the point of reference for dramatic interventions when these reconstructed images clearly excluded indigenous practices of land management. From a scientific standpoint, the flirtation of some restorative ecologists with notions of ritual led critics to wonder if restoration work is not elevated to the status of religious practice. "The restoration ecologist's aim", historian Jack Kirby would write in the 1990s, "is to do nothing less than save neo-European's souls. Salvation is accomplished by ritual, which reconnects human with nature."[66] Strangely enough, the advanced science of engineering ecosystems looped back to the ceremonial functions of trees and forests that were suppressed by modern science for two centuries, cutting across distinctions between "modern" and "traditional" societies in "developed" and "developing" countries found in "tropical" and "temperate" climatic regions. In light of this ceremonial haze, Sonfist's work can be understood as a demonstration of restoration work that acts not as a monument to an ecology, but to the metaphor of ecology.[67] The objections raised by some local residents, who argued that what Sonfist saw as a space of healing forest landscape was, in their view, a dark and dangerous corner, highlight that this metaphor is always at the crux of contestation.

★★★

Some decades after Sonfist installed his primeval environment in Greenwich Village, landscape ecologist Eric Sanderson, working at the Wildlife Conservation Society, published a book summarizing his research into the ecological history of New York City. Sanders led the Mannhatta Project (pun apparently intended) at the Conservation Society for a decade, and the book was intended less as a common ecological guide with botanical illustrations and species succession diagrams, and more as a tour de force of speculative restoration ecology.[68] In his research, Sanderson scrupulously retraced and identified the current locations and altitudes of the 573 hills that gave Mannhatta its name in Munsi, the language of the Lenape who lived on the island when Henry Hudson sailed up the river in 1609. Many of these topographic cues were long gone, overrun by numerous waves of developmental zeal. Therefore, the research project had to find ways to recover earlier layers. Drawing, among other sources, from John Randel Jr.'s well-known maps of the island, surveyed as part of his work with the commissioners in charge of the plan that would eventually lay down the Manhattan Grid in 1811, Sanderson and his team filtered out ownership lines and post-seventeenth-century structures to extract the natural features underneath.[69] They then matched this cartographic information to GPS data collected at key points throughout the city to be able to reconstruct the streams, dips, marshes, and ravines that once composed the island's landform and made it, in the words of Hudson, "as pleasant a land as one can tread upon" (Figure 4.6).[70]

Then began a long process of ecological reconstruction, in which the Mannahatta team essentially poured an inventory of flora and fauna likely to have existed four hundred years prior into different areas of the city. This process was based on scientific facts as it was a work of speculative design: beginning with a list of ecosystems based on parameters such as soil type, rainfall, and the like, the team populated their imaginary territory with species that possibly inhabited it, and gradually sketched what Sanderson termed a Muir Web – a visualization of the network

FIGURE 4.6   Photocollage from the Mannhatta Project, 2009. © Markley
Boyer and Eric W. Sanderson, Wildlife Conservation Society and
New York Botanical Garden/Yann-Arthus Bertrand, CORBIS.

of connections between species, habitats, and ecosystems that cords, in Muir words, "anything … to everything else in the universe". The use of Muirian rhetoric as a structural analogy for ecological work betrays the holistic, even spiritual, undercurrents that flow alongside Mannhatta's analytic methodologies. Eventually, the team compiled a database of more than one thousand species and eight thousand links between species and between species and habitats. These included 24 species of mammals, 233 of birds, 32 of reptiles and amphibians, 85 of fish, 627 of plants, and unknown numbers of fungi, lichens, mosses, insects, shellfish, and other invertebrates. Specifying numbers was important as it balanced the speculative nature of the project, lending it an empirical front appropriate for a scientific endeavor.

After spending a long time immersed in a lost past, Sanderson was undoubtedly experiencing visions of his own. In a *National Geographic* article, he is described as crossing Seventh Avenue and seeing in his mind's eye a swampy creek that "would have been a good place for deer, wood ducks … as well as eels, pickerel and sunfish", just beneath the Marriot Marquis at the corner of Broadway and West 46th street.[71] This direct access to ecological ecstasy pulls Mannhatta out of the deep waters of times past. If the portal to the oak–hickory world is readily available near the gutters of Seventh Avenue, then perhaps the green doppelganger of New York City was both a past and a future. While Sanderson argued that the project is not about going back to the state of the island before colonization, the poetic license of its visualizations says otherwise (Figure 4.6).

These striking images were created by three-dimensional specialist Markley Boyer, who was brought on board to transform the data sets into scenes in which computer software habitually used in the film industry allowed for the modeling and positioning of tens of thousands of trees for each forest type. The impact of before-and-after aesthetics is cleverly played out in the resulting visualizations that are weaved throughout the book: a contemporary photograph of Times Square is placed next to a scene, taken from the same angle, in which a deer and a beaver stand still amid a red maple swamp. Or, elsewhere, a low aerial of Foley Square is contrasted with an Arcadian landscape in which centuries of colonization and pollution are undone to expose a collection pond in the middle of a forest, thinly dotted with Lenape longhouses. The centerpiece is a visualization of the entire island from the south in its primeval state. A lone American eagle glides above the billowy green, with smoke indicating indigenous presence rising not far from Greenwich Village, where Time Landscape is freed from its iron cage and multiplied a thousand times. Like Hugh Ferriss, rendering the metropolitan energy of early twentieth-century New York visible, Boyer articulates the visual aura of a total project of restoration ecology. Through his imaging, Mannhatta emerged as an ecological and spatial design project that would befit the creative imagining Mostafavi was aiming for with Ecological Urbanism, only that the arrow of time was reversed: instead of capitalizing on current crises by planning a speculative future city, Mannhatta concentrated on speculatively designing a primeval past in all the glory of detail supplied by the powers of advanced computing. The early 2000s were saturated with impending climate change, complete with hurricanes razing cities, infrastructures, and agricultural land, fully packed with scientific premonitions of cities going underwater due to the projected rise in sea levels. In the gloom of doom, Mannhatta suddenly appeared as a possible scenario for the era right after the deluge. In this possible future, the spiritual values assigned to trees and forests in faraway continents and faraway times recaptured the heart of New York City, and the ecological forest was placed as a talisman, offering magical protection against the fallacies and hubris built up during four hundred years of island colonization.

Notes

1 Mostafavi, Mohsen, and Gareth Doherty. *Ecological Urbanism*. Cambridge, Mass.; Baden, Switzerland: Harvard University Graduate School of Design, Lars Müller Publishers, 2009, 17.

2 The conference brought together speakers from a variety of disciplines and approaches, including ecologist Richard T. T. Forman, New Urbanist Andrés Duany, William Mitchell of MIT's Smart Cities program, smell expert Sissel Tolaas, and postcolonial theorist Homi K. Bhabha.

3 For a typical description of the "spread" of vertical forest projects, see Lubell, Sam. "Green Grow His Towers." *National Geographic* 239, no. 5 (2021): 22.

4 This particular phrase appears in Boeri's website and press releases and was filtered to numerous news pieces covering the project. See "Vertical Forest: A Sustainable Residential Building." *urbanNext*. Accessed March 5, 2024. https://urbannext.net/vertical-forest/ or Mun-Delsalle, Y-Jean. "Milanese Architect Stefano Boeri Builds Forests in the Sky." *Forbes*, July 1, 2021.

5 Boeri referred to the project as a "comfort-creating 'mechanical' device". Quoted in Barber, Daniel A., and Erin Putalik. "Forest, Tower, City: Rethinking the Green Machine Aesthetic." *Harvard Design Magazine*, no. 45 (March 2018): 234.

6 It is telling that the comparison with the forest is presented in urban design rather than ecological terms: "to place the green spaces on flat land, the necessary amount of land would occupy 10,000 square meters in a natural forest and 50,000 square meters in a residential setting". Grieco, Lauren. "Stefano Boeri:

Bosco Verticale/Vertical Forest in Milan." *Designboom*, November 10, 2011. https://www.designboom.com/architecture/stefano-boeri-vertical-forest.

7 Tabb, Phillip, and A. Senem Deviren. *The Greening of Architecture: A Critical History and Survey of Contemporary Sustainable Architecture and Urban Design.* Farnham, Surrey, England; Burlington, VT: Ashgate, 2013.

8 Mostafavi and Doherty, *Ecological Urbanism*, 122. Also quoted by Tabb and Deviren.

9 Wines, James. *Green Architecture*. Köln; New York: Taschen, 2000, 20.

10 Wines, James. *Green Architecture*. Köln; New York: Taschen, 2000, 11.

11 SITE worked with landscape experts to be able to keep as much as the trees and soil as possible. Belogolovsky, Vladimir. "Architecture needed to be liberated from itself." August 30, 2022. *STIRworld*. Accessed March 6, 2024. https://www.stirworld.com/think-columns-architecture-needed-to-be-liberated-from-itself-says-james-wines-of-site.

12 This artistic circle included, beyond Matta-Clark, Michael Heizer, Robert Smithson, Vito Acconci, Alice Aycock, Hannah Wilke, Nancy Holt, and Alan Sonfist. Belogolovsky, Vladimir. "Architecture needed to be liberated from itself." August 30, 2022. *STIRworld*. Accessed March 6, 2024. https://www.stirworld.com/think-columns-architecture-needed-to-be-liberated-from-itself-says-james-wines-of-site.

13 The logo treatment presented another point of divergence from Denise Scott Brown and Robert Venturi's earlier critique of modernist architectures, which prioritized signs and communication over volume and form. See Scott Brown, Denise. "Learning from Pop." *Casabella*, December 1971, 359–60.

14 Restany, Pierre, Bruno Zevi, and SITE, Inc. *SITE – Architecture as Art*. London: Academy Editions, 1980, 31.

15 McGraw, Patrick. "Must What Went Up, Come Down? An Interview with SITE's James Wines." 032c, December 17, 2020. https://032c.com/magazine/should-what-went-up-come-down-an-interview-with-sites-james-wines. Accessed March 6, 2024.

16 Only years later did the forest building enter the conversations of architectural critique as a forerunner of a different, more environmentally sensitive, kind of architecture. See Dean, Penelope. "Never Mind All That Environmental Rubbish, Get On with Your Architecture." *Architectural Design* 79, no. 3 (2009): 24–9. Even so, a contemporary critic maintained that it was "cute and clever, but hardly germane to a serious conversation about environmental or ecological architecture". See Freeman, Belmont. "Aesthetic Environmentalism." *Places Journal*, November 2023. https://placesjournal.org/article/aesthetic-environmentalism-review-emerging-ecologies-moma-exhibition/. Accessed March 6, 2024.

17 As he was pushing for green architecture, Ambasz designed "several fuel engines for the decidedly nongreen Cummins Engine Co., where he has been a consultant for 29 years". LaBarre, Suzanne. "The Mystery and Mythology of Architect Emilio Ambasz." September 1, 2009. *Metropolis*. Accessed March 7, 2024. https://metropolismag.com/projects/mystery-mythology-architect-emilio-ambasz/.

18 Eisenman, the most prolific provocateur of this cadre, established the genre of architecture informed by linguistics. His participation in the meeting of the CASE (Conference of Architects for the Study of the Environment) group organized by MoMA Director of Architecture and Design Arthur Drexler in 1969, led to his inclusion in the milestone *Five Architects* publication and to several built commissions that further amplified the impact of his theories. His intellectual influence lasted well into the 1990s with a new generation of architectural theorists. See Museum of Modern Art. *Five Architects: Eisenman Graves Gwathmey Hejduk Meier.* New York: Wittenborn & Company, 1972. For an introduction to the architectural interest in semiotics see Agrest, Diana, and Mario Gandelsonas. "Semiotics and Architecture: Ideological Consumption or Theoretical Work." *Oppositions* 1 (1973): 93–100.

19 Robert A. M. Stern wrote on this thread of architecture: "[its] buildings are very much of a time and place: they are not intended as ideal constructs of perfected order; they select from the past in order to comment on the present". See Stern, Robert A. M. "Gray Architecture as Post-Modernism or Up and Down from Orthodoxy." In *Architecture Theory since 1968*, edited by K. Michael Hays, 242–5. Cambridge, Mass.: The MIT Press, 1998 [1976].

20 Mays, Vernon. "The Elusive Mr. Ambasz." *Architect (Washington, D.C.)* 98, no. 6 (June 1, 2009): 64.

21 The Museum of Modern Art. "Emilio Ambasz/Steven Holl: Architecture. February 9–April 4, 1989." February 1989. https://assets.moma.org/documents/moma_press-release_327533.pdf. Accessed March 7, 2024.

22 A 2000 thermal study by planning partner Takenaka Corporation with the Nippon Institute of Technology and Kyushu University demonstrated the green roof's impact in alleviating the heat island phenomenon. See Velazquez, Linda. "Greenroofs Project of the Week: ACROS Fukuoka Prefectural International Hall." August 12, 2011. https://www.greenroofs.com/2011/08/12/gpw-acros-fukuoka-prefectural-international-hall/. Accessed March 7, 2024.

23 Ambasz reflected that "by 'green over gray,' I meant the replacement of the notion of architecture as gray and the integration of architecture, which is man-made, with natural nature, which in some cases I've achieved." See Davidson, Cynthia. "The Environments of Emilio Ambasz." *Log*, no. 52 (summer 2021), 101.

24 Even Frank Lloyd Wright, who spent a lifetime making an argument for a kinship between architecture and land never went deeper than the horizontal surface, which was in turn an inheritance from the Jeffersonian grid that parceled the entire American landmass in accordance with the bylaws of two-dimensional geometry and two-dimensional thinking. The modernists that followed Wright in the twentieth century were perfectly at ease with an even higher vantage point and minimal contact with the ground in their schemes for mass housing in Frankfurt or Brasília and new towns in Le Havre or Algiers.

25 Yeang, Kenneth King-Mun. "Theoretical Framework for Incorporating Ecological Considerations in the Design and Planning of the Built Environment." PhD diss., University of Cambridge, 1980.

26 These include considering sites in environmental terms, focusing on energy and material flows, and taking into account the life cycles of various designed components. See Yeang, Ken. *Designing with Nature: The Ecological Basis for Architectural Design.* New York: McGraw-Hill. 1995, 34.

27 Yeang, Ken. *Designing with Nature: The Ecological Basis for Architectural Design.* New York: McGraw-Hill. 1995, 187.

28 Yeang, Ken. "Designing the Tropical Skyscraper." In *Mimar 42: Architecture in Development*, edited by Hasan-Uddin Khan, Concept Media Ltd., 1992, 41.

29 Yeang, Ken. "Designing the Tropical Skyscraper." In *Mimar 42: Architecture in Development*, edited by Hasan-Uddin Khan, Concept Media Ltd., 1992, 43.

30 In later decades, Yeang emphasized the ecological aspects of his oeuvre, but the works from the early period demonstrate his interest in contemporary architectural styles such as deconstructivist and high-tech architecture.

31 Jorgensen, Erik "The History of Urban Forestry in Canada." In *The Proceedings of the First Canadian Urban Forest Conference*, Winnipeg, MB, May 30 to June 2, 1993, 17.

32 Jorgensen, Erik. "Urban Forestry in the Rearview Mirror." *Arboricultural Journal* 10, no. 3 (1986): 177.

33 Konijnendijk, Cecil C., Robert M. Ricard, Andy Kenney, and Thomas B. Randrup. "Defining Urban Forestry – A Comparative Perspective of North America and Europe." *Urban Forestry & Urban Greening* 4, no. 3–4 (2006): 94.

34 Jonnes, Jill. *Urban Forests: A Natural History of Trees and People in the American Cityscape.* New York, New York: Viking, 2016, ch. 12, Kindle.

35 The act posited that "urban trees are 15 times more effective than forest trees at reducing the buildup of carbon dioxide and aid in promoting energy conservation through mitigation of the heat island effect in urban areas" and at the same time that they can "contribute to the social well-being and promote a sense of community". See US Congress, *An Act to Authorize the Secretary of Agriculture to Provide Cooperative Forestry Assistance to States and Others, and for Other Purposes*, Pub. L. 95–313, §1, 92 Stat. 365 (July 1, 1978).

36 US Congress, *An Act to Authorize the Secretary of Agriculture to Provide Cooperative Forestry Assistance to States and Others, and for Other Purposes*, Pub. L. 95–313, §1, 92 Stat. 365 (July 1, 1978).

37 Reynolds, Harris A. "An Opportunity for the Young Landscape Architect." *Landscape Architecture* 4, no. 2 (1914): 47–51.

38 Beatty participated in the first national urban forestry conference in Washington, D.C., in 1978 and taught the first urban forestry course at UC Berkeley. See Beatty, Russell A. "Openings in Urban Forestry." *Landscape Architecture* 70, no. 6 (1980): 598.

39 Beatty published an early survey of tree planting programs in seventy-two American cities, in which both quantitative and qualitative data were recorded. Working on an urban tree plan in Lafayette, California, he used its publication to call landscape architects to "demonstrate how the individual pieces of urban forestry can be shaped into a comprehensive whole through systematic planning". See Beatty, Russell A., and Craig T. Heckman. "Survey of Urban Tree Programs in the United States." *Urban Ecology* 5, no. 2 (November

1981): 81–102, and Beatty, Russell A. "Planning the Urban Forest." *Landscape Architecture Magazine* 71, no. 4 (1981): 458.

40 Jonnes, *Urban Forests*, ch. 12, Kindle.

41 Many US cities experienced financial difficulties in the 1970s, which required a solid economic argument for urban trees.

42 Rowntree's collaboration with Fred Bartenstein, special assistant to the city manager of Dayton, is described at length in Jonnes's book on the history of urban forests. Jonnes, *Urban Forests*, ch. 12, Kindle.

43 Rowntree's definition is based on the definition of a forest is that at least 10% of the land is stocked with trees. See Rowntree, Rowan A. "Ecology of the Urban Forest – Introduction to Part I." *Urban Ecology* 8, no. 1–2 (1984): 2.

44 The differences between the environmental characteristics of buildings and trees are fundamental and numerous, and yet it was argued that the bene-fits of urban meteorology and climatology outweigh the shortcomings of this comparison. See Oke, T. R., J. M. Crowther, K. G. McNaughton, J. L. Monteith, and B. Gardiner. "The Micrometeorology of the Urban Forest [and Discussion]." *Philosophical Transactions of the Royal Society of London. Series B, Biological Sciences* 324, no. 1223 (1989): 335.

45 Kaplan, Tracey. "Ambitious Plan to Preserve 'Urban Forest': Thousand Oaks Counts Its Trees to Save Them." *Los Angeles Times*, June 25, 1989.

46 Konijnendijk et al., "Defining Urban Forestry", 97.

47 Shen, Guofang. "Three Decades of Urban Forestry in China." *Urban Forestry & Urban Greening* 82 (2023): 127877.

48 The cost-effectiveness of this plan was later put to question. See Escobedo, Francisco J., John E. Wagner, David J. Nowak, et al. "Analyzing the Cost Effectiveness of Santiago, Chile's Policy of Using Urban Forests to Improve Air Quality." *Journal of Environmental Management* 86, no. 1 (2008): 148; CONAMA. "Plan de Prevención y Descontaminación Atmosférica." Comisión Nacional del Medio Ambiente, Santiago, Chile, 1997.

49 Stoffberg, Hennie, Margaretha van Rooyen, Johan van der Linde, et al. "Carbon Sequestration Estimates of Indigenous Street Trees in the City of Tshwane, South Africa." *Urban Forestry & Urban Greening* 9, no. 1 (2010): 9–14.

50 Borelli, Simone, Michela Conigliaro, and Federica Di Cagno. *Urban Forests: A Global Perspective*. Rome: FAO, 2023, 210. Accessed March 18, 2024. https://doi.org/10.4060/cc8216en. One important exception to the trajectory of urban forestry into the less temperate, less affluent regions is Singapore, in which greening schemes existed since the 1960s, allowing a smooth transition into the urban forest era.

51 The rates were similar in all urban samples, ruling out the option of local anom-alies. However, all were much lower than in rural samples. White, Carleton S., and Mark J. McDonnell. "Nitrogen Cycling Processes and Soil Characteristics in an Urban versus Rural Forest." *Biogeochemistry* 5, no. 2 (1988): 243.

52 A 2015 meta-analysis of more than five hundred sources from 1988–2014 found an overwhelming percentage of contributions from North American and (after 2000) European authors. See Ostoić, Silvija Krajter, and Cecil C.

Konijnendijk van den Bosch. "Exploring Global Scientific Discourses on Urban Forestry." *Urban Forestry & Urban Greening* 14, no. 1 (2015): 129. For a glimpse into the disparities between different definitions of urban forestry, see Konijnendijk et al., "Defining Urban Forestry", 100.

53 One example can be found in the previously mentioned article on the discourses of urban forestry, which opens by arguing that "broad consensus exists on the multiple benefits urban forests provide accompanied with growing body of scientific evidence". It then cites another article as a source, which in turn discusses the incongruities of tree benefit assessment and calls for additional research to be conducted. See Konijnendijk et al., "Defining Urban Forestry", and Roy, Sudipto, Jason Byrne, and Catherine Pickering. "A Systematic Quantitative Review of Urban Tree Benefits, Costs, and Assessment Methods Across Cities in Different Climatic Zones." *Urban Forestry & Urban Greening* 11, no. 4 (2012): 351–63.

54 A 2022 study identified 181 tree planting projects in fifty-four cities covering all African countries except for Guinea-Bissau. The FAO urban forestry report mentions the spiritual values associated with some of these projects, noting, "This situation contrasts with findings in the Global North, where spiritual values have been ranked lowest in importance compared to other cultural services such as recreation." See Lobe Ekamby, Emmanuel S. H., and Pierpaolo Mudu. "How Many Trees Are Planted in African Cities? Expectations of and Challenges to Planning Considering Current Tree Planting Projects" *Urban Science* 6, no. 3 (2022): 59. https://doi.org/10.3390/urbansci6030059. And Borelli, Simone, Michela Conigliaro, and Federica Di Cagno, *Urban Forests: A Global Perspective*, 178.

55 Sonfist describes the formation of an early kinship with the trees, developed during family trips to the forest and influencing his concept: "My dreams of my childhood tree families still are my dreams. My childhood memories of families of trees became my blueprint for the Time Landscape." Handel, Dan. Interview with Alan Sonfist. [Email interview]. April 5, 2024.

56 Sonfist, Alan. "Natural Phenomena as Public Monuments." Presented at the Metropolitan Museum of Art in 1968. In *Theories and Documents of Contemporary Art: A Sourcebook of Artists' Writings*, edited by Peter Selz and Kristine Stiles. 2nd ed., revised and expanded by Kristine Stiles. Berkeley [Calif.]: University of California Press, 2012, 624–5.

57 Sonfist went into researching old maps of the island in the library system, educating himself on the geology and botany of New York, and creating a palette of native trees to be planted.

58 Sonfist worked for a while with the vice-director of the MET, architect Arthur Rosenblat, to develop a plan for the American wing. Handel, Dan. Interview with Alan Sonfist.

59 Sonfist worked for a while with the vice-director of the MET, architect Arthur Rosenblat, to develop a plan for the American wing. Handel, Dan. Interview with Alan Sonfist.

60 Perlberg, Deborah. "Alan Sonfist, Laguardia Place." *Artforum* 17, no. 1 (September 1978): 84.

61 Pollan, Michael. *Second Nature: A Gardener's Education.* New York, NY: Grove Press, 1991, 113.

62 Sonfist notes that in the 1960s, architects and city planners referred to many indigenous plants as "weeds", highlighting the cultural transformation of the term.

63 Martin, Laura J. *Wild by Design: The Rise of Ecological Restoration.* Cambridge, MA: Harvard University Press, 2022, 7.

64 Jordan, William. "Making Nature Whole: Fifty Years of Ecosystem Restoration at the University of Wisconsin Arboretum." *The George Wright Forum* 2, no. 4 (1982): 36. Jordan, an ecologist working at the University of Wisconsin's arboretum, was a key figure in the definition of restoration ecology as a distinct subdiscipline.

65 Significantly, this process required a change in ecological thinking, *"expanding the category of 'environmental disturbance' beyond windstorms, fires, and floods to include nuclear bombs – and, ultimately, other human actions like deforestation and pollution".* In turn, this facilitated the introduction of a constant threat into the theory of ecosystems. Martin, *Wild by Design,* 115.

66 Kirby, Jack. "Gardening with J. Crew: The Political Economy of Restoration Ecology." In *Beyond Preservation: Restoring and Inventing Landscapes,* edited by A. Dwight Baldwin, J. De Luce, and C. Pletsch. Minneapolis: University of Minnesota Press, 1994, 238.

67 As can be seen in Sonfist's ambition to situate his forest within the long-gone indigenous environment, "near a trout stream, and the original native trail, Broadway". Handel, Dan. Interview with Alan Sonfist.

68 Sanderson, Eric W., and Wildlife Conservation Society. *Mannahatta: A Natural History of New York City.* New York: Abrams, 2009.

69 Randel was hired by the commission to survey the island, a task he carried on with for over a decade to produce ninety-two sheets at a scale of 100 feet to 1 inch, the most detailed maps of any American city up to that point. The results of his meticulous work, produced in collaboration with his wife Matilda and known as the Randel Farm maps, were published in 1820. Sanderson later worked on the earlier British Headquarters Map of the island from 1782. See Sanderson, Eric W., and Marianne Brown. "Mannahatta: An Ecological First Look at the Manhattan Landscape Prior to Henry Hudson." *Northeastern Naturalist* 14, no. 4 (2007): 545–70.

70 The entry from Hudson's journal was included in Johannes de Laet's *New World,* published in Dutch in 1625. Translation included in Jameson, J. Franklin. *Narratives of New Netherland, 1609–1664.* New York: C. Scribner's Sons, 1909, 49.

71 Miller, Peter. "Before New York." September 01, 2009. *National Geographic.* Accessed March 27, 2024. https://www.nationalgeographic.com/magazine/article/manhattan.

5

UBIQUITOUS INTELLIGENCE

In 2016, researchers working at the Berlin University of the Arts published a white paper proposing a prototype for a forest that can own and sustain itself. The three, artists Paul Seidler and Paul Kolling, joined by developer and crypto researcher Max Hampshire, argued that by augmenting a forest tract with sensing apparatuses and blockchain technology, it would be able to negotiate its log sales through smart contracts, accumulate capital, and thus achieve economic autonomy. At this point, the forest would be free to expand by purchasing more land for future development. A forest making transactions on its own may sound unnerving, but the authors posited that terra0 simply builds on a legal grey area in which "Blockchain technology and smart contracts enable non-human actors to administer capital and therefore to claim the right to property for the first time". Arguing for the inclusion of nonhuman actors in the economic realm, they write: "Since an individual's property is protected in accordance with their rights, one would assume that objects which have gained the right to property are entitled to similar personal rights as natural persons" (Figure 5.1).[1]

This argument builds on developments in environmental law and resource management that gained traction around the time of writing. In their paper, Seidler, Kolling, and Hampshire refer to two well-publicized cases: the introduction of the Rights of Nature article in the constitution of Ecuador in 2008 and the granting of legal personality to the Whanganui river the New Zealand, unfolding for almost a decade until its introduction as a public act in 2017.[2] These examples were signaling a very active scene: in 2011 river systems in the state of Victoria, Australia, were offered protection under the concept of legal rights to nature. In 2016, the Constitutional Court of Colombia asserted that the Atrato River had rights. In 2017, the Ganges and Yamuna rivers joined the current as the high court of the state of Uttarakhand, India, declared them as possessing "all of the rights, duties, and liabilities of a living person".[3] While certain

DOI: 10.4324/9781003473411-6

FIGURE 5.1   Can an augmented forest own and utilize itself? Terra0, 2016.
© terra0, https://terra0.org.

legal scholars doubt the efficacy of such legislation, it is becoming widely
recognized by governments and policymakers as a viable course of action
in addressing long-standing land disputes involving indigenous populations.[4] The conversations surrounding the rights of non-human entities
seem to embrace a more holistic approach towards the environment,
sometimes associated with non-Western thinking. However, its effectiveness in court is deeply rooted in legal theory and environmental legislation
as developed in the American system. In 1972, Christopher D. Stone, law
professor at the University of Southern California, published a long paper
titled "Should Trees Have a Standing? – Toward Legal Rights for Natural
Objects", which would become a seminal text for environmental law and
environmental thinking. In it, he established that trees and other natural
elements indeed have a standing in court, that is, that they can be represented and bring suit, Stone brilliantly describes the evolution of granting
legal rights as one that progresses by leaping into the "unthinkable": at
one time in the not-so-distant past, he writes, children, women, enslaved
people, and Native Americans did not possess rights in court, and "each
successive extension of rights to some new entity has been, theretofore, a
bit unthinkable."[5] This evolution was not restricted to humans. Dwelling
on the theoretical difficulties facing medieval legal scholars when defining
the rights of the church or of kingdoms as extending beyond those of their

human representatives, and the similar conundrums facing judges in the nineteenth century when addressing "corporate bodies", Stone reminds us that inanimate right-holding entities such as "trusts, corporations, joint ventures, municipalities ... partnerships, and nation-states" fill the world of lawyers and establish the ground rules according to which the standing of natural entities could be considered. Each time a proposition is made to confer rights onto a new entity, it may sound odd or counterintuitive. That is since "until the rightless thing receives its rights, we cannot see it as anything but a thing for the use of 'us' – those who are holding rights at the time".[6] Stone's proposition for the inclusion of the rights of natural entities is based on the guardianship approach, essentially securing an effective voice for the environment through its representation by an expert organization possessing the required technical and legal expertise. He then discusses in detail the legal and economic ramifications of this concept.[7] Only later in his paper does Stone get to the deeper changes in consciousness required for the leap beyond homocentric concerns. Curiously quoting at length from occult spiritualist and astrologer Dane Rudhyar, whose writings envisioned the emergence of transpersonal planetary consciousness, Stone connects the day-to-day workings of legal systems to the "conscious mind of the earth". He then concludes with

> as radical as such a consciousness may sound today, all the dominant changes we see about us point in its direction.... Increasingly, the death that occupies each human's imagination is not his own, but that of the entire life cycle of the planet earth, to which each of us is as but a cell to a body.[8]

Forests had been considered in human terms before, even in the most unlikely places. Gifford Pinchot, writing his well-circulated *A Primer of Forestry* – supposedly a technical manual on the proper management of forests – described in it the life of the forest as the history of "the help and harm which the trees receive from one another"[9] and their coexistence and struggle with and against each other. Under the veneer of a rational-scientific exposition, he persistently politicized this "community of trees", to be portrayed at one point as "the inhabitants of a town",[10] at a second as striving "for the good things in life",[11] and in a third as one whose subjects are comforting and assisting "the other trees, which are [their] friendly enemies".[12] Some decades later, Aldo Leopold argued that if human and forest communities could be regarded in similar terms, then the ethics that arise from membership in a human community could be extended to include the nonhuman world.[13] Once this door was opened, it equipped environmentally minded judges with the arguments necessary to protect the rights of non-humans. As did Supreme Court judge William Douglas in 1972, writing that "environmental issues should be tendered by the inanimate object itself".[14] Referring to Leopold's Land Ethic and Stone's text, the judge added: "[S]o it should be as respects valleys, alpine meadows, rivers, lakes, estuaries, beaches, ridges, groves of trees, swampland, or even air that feels the destructive pressures of modern technology and modern life."[15] It was precisely through such rulings that Stone's argument acquired much influence, eventually coalescing with the views and advocacy of indigenous groups outside of the United States to result in pioneering acts of granting rights to nature.[16]

Beyond questions of standing, this still unfolding legal history obliquely cements the idea of natural elements endowed with a consciousness. This concept is often associated with the rise of deep ecology – the banner that allied ecologists, philosophers, and historians beginning in the 1970s around the critique of anthropocentric worldviews, industrial development, and Western science. These thinkers made frequent sorties into spiritual and metaphysical arguments in their attempts to strike the right balance between environmental ethics and ecological metaphysics.[17] Yet, the rise of the specific metaphor of forest intelligence, which is a subset of the more-than-human consciousness proposition, is in fact closely linked with the emergence of the idea of machine intelligence.

★★★

Assigning intelligence to inanimate entities is commonplace to the point of neutrality in contemporary technological societies. Whether the case in point is an autonomous vehicle or a natural system, there will be ample evidence and enough witnesses ready to vouch for their superior sagacity and promptness to bypass the fallacies and imperfections of humans. In part, relying on machines is a reaction to the unfolding environmental and financial crises of the last few decades, but it also aligns with a latent desire among the technological class to organize the world according to its own value systems. The idea has been circulating in scientific circles for decades: if inapt politicians can be held in check by a total rationalization of decision-making systems, and if the desires of the general populace can be engineered to follow predetermined routes, then the true potential of progress could be unleashed in both "developed" and "underdeveloped" countries. Starting at that time, scientists attempted to apply game theory and systems theory to the global arena, rationalizing the workings of leaders and nations to build future scenarios.[18] Following this line of thinking, in the mid-1960s, R. Buckminster Fuller proposed the World Game as a way to apply a holistic design approach to global challenges.[19] Players attempted to address issues of resource availability and respond to population trends and needs constantly fed by a computer while advancing collectively to solve problems on a room-sized map. The game was meant to steer Spaceship Earth above and beyond what Fuller regarded as the "illogic of 200 nation state admirals".[20] All that was needed was a global system of governance and full trust in the workings of design science. If the idea that a small group of predominantly white graduate students at Stanford or the New York School of Painting and Sculpture could make informed decisions regarding the immensely complex, multi-layered challenges facing remote peoples in remote places that have never visited may seem naïve, adherents of the World Game argued that its shortcomings were quantitative and not qualitative. If enough information could be fed into the computer and presented to players in real time, the game would be able to achieve its goals. And, goes the reasoning, if the human players reach their capacity of processing information, they can be aided, or eventually substituted, by intelligent machines.[21] This conviction was strong enough to develop spin-off games that would focus on sustainable development and global food distribution.[22] From the outset and throughout a decade of experimentation, Fuller's intentions of collapsing the boundaries between human and machine intelligence were clear: "The human mind invented the computer as an extension of humanity's integral computer, information storing and retrieving system, the brain." The game was the vehicle to get there: "our World Game", he wrote, "will be in effect a World Brain."[23]

And so, machine intelligence was established in design. But as intelligence is not an observable fact but an assumed quality, it often rested on nebulous definitions. In Alan Turing's view, as expressed by his 1950 "imitation game", the point was not so much whether a machine was intelligent, but rather whether it was able to display intelligent behavior to the degree that it would be indistinguishable from human behavior. Turing famously introduced his thought experiment by writing: "I propose to consider the question: 'can machines think'?"[24] And while the Turing Test would later be criticized for its incapacity to indicate intelligence or consciousness in machines, it remains Turing's most influential contribution, inspiring a variety of well-known philosophical inquiries and computer programs.[25] The postulation of thinking machines was put forth in a feverish period for cybernetics, in which experiments with stroboscopic light and ideas of synthetic brains broke through disciplinary boundaries and permeated works of architects, musicians, engineers, corporate managers, and designers.[26]

In 1964, the International Business Machines company recruited J. C. R. Licklider to work in its Yorktown Heights headquarters as manager of information sciences, systems, and applications. The somewhat obscure title indicated the company's struggle to define Licklider's trailblazing work in diverse fields such as acoustics, network communication, and information processing. Trained in psychology and psychoacoustics, "Lick" unassumingly slid into the role of interactive computing prophet. In 1960, he published *Man-Computer Symbiosis*, in which he anticipated the coming of a close cooperation between men and electronic computers.[27] Comparing this relationship to the interdependence between a fig tree and its pollinator insect *Blastophaga grossorun*, which lives in its ovary, Licklider implicitly advanced the understanding of computers as organisms, a metaphor that will gather traction for decades all the way to the current age of cognitive computing. "Human brains and computing machines", he writes, "will be coupled together very tightly."[28] The resulting symbiosis will "think as no human brain has ever thought and process data in a way not approached by the information-handling machines we know today".[29] Licklider's symbiosis was not only a metaphor but an allegory. As clear as the scientific fact that insect and tree cannot live without each other, it implied that humans cannot survive the speed and information load of the modern world without the aid their mechanical partner.

Once computing machines could be considered symbiotic with biological brains, they were no longer external to the workings of the human mind. IBM seemingly seized on this notion as it began a decades-long journey of anthropomorphizing its products. Computers were no longer following coded instructions but "thinking", "understanding", and "solving problems". One important highlight in this campaign was the IBM Pavilion at the 1964 New York World's Fair, designed by Charles and Ray Eames in collaboration with Eero Saarinen and Associates. The IBM pavilion was a centerpiece of the collective fantasy of the fair, where corporate experience and advertising were skillfully stirred together to present a shiny vision of American omnipotence. On the sales front, IBM's information machines were making quantum leaps into the mainframe era, and the frontiers of mass markets seemed promising. The designers involved were already acting as efficient boosters of the company's vision of ubiquitous intelligence. Eero Saarinen was the ultimate form-giver of corporate culture in the 1950s and early 1960s designing, among other things, the signature tent for the Aspen design conferences. After his passing, his firm continued its engagement with executive dream spaces under Kevin Roche, who oversaw the design of IBM Pavilion. The Eames duo, from their side, had been making films for the company since the late 1950s and proved masterful in communicating corporate messaging with aesthetic and material eloquence (Figure 5.2).

FIGURE 5.2   Postcard of the IBM Pavilion at the New York's World Expo, 1964.

The grounds of the World's Fair in Queens were densely populated by pavilions and attractions aggressively competing for the attention of visitors. Yet, even in between the gangs of moving and talking Disney dummies, the aura of Billie Graham's evangelical film, emanating from his theater, and Michelangelo's *Pietà*, brought in by the Vatican for the occasion, the IBM pavilion did not fail to make a splash. The *New York Times* called it a sensation, *Time* magazine applauded its humor and sophistication, the New York Chapter of the American Institute of Architects cited it for excellence in design, and Esquire identified it as the one distinguished exhibit that is a notable exception to its preceding jaundiced report on the "spectacle on Flushing Meadow".[30] Ada Louise Huxtable, highly critical of the entire fair enterprise, described the IBM contribution as "an imaginative funfair of information … [that] proves that the corporate message can be put across as an integrated architectural-design concept, and without Walt Disney."[31] The exhibit was a "54,000 square foot display dominated by a 90-foot-high egg nesting in a sprawling grove of steel trees".[32] It was an unusual building by all accounts. The open ground floor under the black trees contained five kiosks, each dedicated to a didactic demonstration of machine thinking. The Eames Studio designed these kiosks in sharp contrast with their contents to resemble a nineteenth-century fair – made of enamel on metal and held by solid wood polls decorated with metal flags and brass finishing. Among these relics of traditional architecture were the slick contemporary boxes of state-of-the-art company products performing all sorts of magic tricks. The Probability Machine exhibit was a demonstration of bell curves with 17,000 polyethylene balls, falling into place every 14 to 18 minutes. Elsewhere, two IBM 1050 machines exhibited a

Russian-to-English translation process by sending and receiving data in almost real time to the company's facility in Kingston, New York. In the character-recognition area, people could handwrite dates, and the machine would retrieve news headlines from the same date and use an IBM 1460 system to print them out. Visitors would then be invited to take their seats on the "people wall" – a hydraulic amphitheater that would shoot them up, four hundred at a time, into the looming ovoid object, decorated with a pattern of embossed IBM logos. Within this space, dubbed by the company "The Information Machine", visitors were bombarded for 10 intense minutes with countless images, projected simultaneously on fourteen large and eight small screens, to visualize the thinking processes of human and computer minds. The immersive trip into the world of information technology in which the audience was wrapped by images, was applauded in both real time and retrospect, acting as a poster project for what architectural historian Beatriz Colomina called "multimedia architecture", in which "the speed of the film is meant to be the speed of the mind".[33]

Most accounts were understandably distracted by the innovative use of images and sound, overlooking the pavilion's architecture, which was based on the decision to represent the new world of machine intelligence through the form of a forest. Within the Saarinen firm, Roche, who took over the design, proposed the steel forest made of columns with hexagon-shaped tops to support the plane above which the theater was to float.[34] The documents from the firm's archives show a pulsating design development process, in which the 32-foot-high pillars that hold the translucent gray-and-green plexiglass roof oscillate between naturalistic to abstract geometries and between wood to steel as the material of choice. One drawing ventures into art nouveau territories, showing a column adorned with hundreds of hand-drawn green leaves. Others explore tree-like morphologies and canopy compositions that attempt to strike the right balance between forest form and machine content (Figure 5.3).

The choice of the forest as the driving metaphor in the pavilion is telling as it is unexpected. In the early 1960s, IBM was expanding rapidly and controlling 70 percent of the US computer market. In 1964, two weeks before the pavilion opened, it launched its high-risk business move of the IBM System/360 – a modular, mainframe family of computers that separated software from hardware for the first time and thus challenged the boundaries between scientific and commercial computers.[35] As the company was in the process of appealing to the mass market, demystifying the workings of intelligent machines became paramount. The Eames astutely expressed this corporate desire, writing in the pavilion's brochure: "we want to dissolve some of the mystery that makes computers seem remote and strange, and show how these complex machines are based on some

FIGURE 5.3   IBM Pavilion metal tree model variations, Kevin Roche, c. 1964.
© Kevin Roche John Dinkeloo and Associates Records (MS
1884), Manuscripts and Archives, Yale University Library.

rather simple problem-solving methods, similar to those people use in
making decisions in everyday life."[36] The naturalization of artificial thinking processes was enveloped by a space articulated using naturalistic cues,
calculatingly placed to smooth the transition into the heart of artifice. The
Eames seductively set the scene:

> follow the winding pathways in this man-made forest to the in-
> triguing world of computers – and then on to the People Wall and

the Information Machine, the huge elevated theater that appears to float on a green canopy high above the grove of trees.

But there was something else. Up to that point IBM, as its omnipresent designer Eliot Noyes expressed it, was in the business of "help[ing] man extend his control over his environment" through the gathering, organizing, and redistributing of information.[37] However the pavilion signaled a parallel, softer approach toward what that environment may be, proposing diffusion rather than seclusion. In this new idea, the naturalistic environment saturated with machines was to play a profound role.

At that time, the environment itself was pulled out of neutrality to become a topic in a heated debate among designers, reaching a cataclysmic confrontation during the 1970 International Design Conference in Aspen (IDCA), Colorado. Initiated in 1949, a thematic conference was first held in 1951 under the heading "Design as a Function of Management". For almost two decades the conferences ran as "a week of top-flight lectures and all-star panel discussions", true to its charter of integrating design in business by exchanging ideas and discussing shared values.[38] Eliot Noyes played an important role in fostering this productive exchange via his responsibilities in the organization of the conference. However, the 1970 edition, "Environment by Design", was destined to be different. The IDCA did deal with "Environment" in one of its previous sessions, being then understood as related to man-made environments.[39] But at this edition, in the heat of protests against the, ecocide in Vietnam, Earth Day consciousness, and environmental legislation, the term gathered different meanings. As noted in another chapter, ecologists like Robert MacArthur were using cybernetic ideas to describe biological environments as complex systems of energy input and output, and IBM was beginning to discuss its interconnected machines in similar terms. The description of environments as systems, becoming more frequent in various professional disciplines, already projected an idea of their inherent logic, aligning with the assumptions of the intelligent machine hypothesis and opening avenues for their design. Perhaps the IDCA organizers had something similar in mind, aspiring to bring what was previously called "nature" under the auspices of managerial design logic. Whatever the intentions were, the event soon went out of hand. The thematic focus on the environment acted like a lightning rod for self-proclaimed radicals wishing to make a claim on the subject. Some were invited; others crashed the convention. Toward its opening the Berkeley troops were clearly gaining ground, and following a chaotic week in the high meadow, the IDCA board of directors was discussing the possibility of discontinuing the conference.[40] The stalemate, wrote Reyner Banham, was resulting from a gulf between those who "believed that rational action was possible within 'the system' and those who wanted out".[41] The crisis was both technical and ideological: if the design of natural systems was tainted by complicity with the existing order, the question was not whether nature could be designed but whether it should be designed at all.

After Aspen, it was becoming difficult for informed designers to treat a given natural environment simply as a set for their follies. Literary critic Leo Marx, writing in 1972 for a volume edited by György Kepes, described Nature as a "transmitter of signals and a dictator of choices", which "defines certain precise limits to human behavior".[42] In many ways, this was an apt description for the unrealized Generator project by Cedric Price, conceived for a wooded site in Florida and designed between 1976 and 1979 (Figure 5.4).

The client was Howard Gilman of Gilman Paper Co., and the site was his company's 7,500-acre White Oak plantation just south of St. Mary's River, which forms the Georgia–Florida state line. Every visionary needs a millionaire, and Gilman, whose company was at one point the largest privately held paper company in the United States, proved eccentric enough for the task. At the time, he was busy quickly amassing an impressive collection of drawings by radical architects, which could serve as an initial point of conversation with Price.[43] The architect and the heir were introduced to each other by art collector and MoMA trustee Barbara Jakobson, who would play a crucial role in the communication and development of the project. Gilman was dreaming of a retreat amid the plantation, one that would propagate creative collaborations rather than lumber for paper towels. Trying to account for the role the forest site would play in the project, Gilman gave Price a cryptic project brief, stating that what was required was "a building which will not contradict, but enhance, the feeling of being in the middle of nowhere", one that "has to respect the wildness of the environment while accommodating a grand piano."[44]

Price delighted in these instructions and responded by proposing what he claimed to be the first truly intelligent building in the history of modern architecture.[45] Generator was imagined as a constantly changing architectural ensemble, made of modular cubical units that could be recombined in real-time by an onsite crane in response to the whims and desires of Gillman's guests. A number of photographs show the initial mock-ups of these units, placed in a clearing on site, with Jakobson seated inside for human scale (Figure 5.5).[46] To further develop this idea, Price invited John and Julia Frazer, whom he knew from professional circles in Cambridge and London, to act as computational consultants, which immediately immersed the project in deep cybernetic waters. The modules were now envisioned as embedded with a single-chip microprocessor

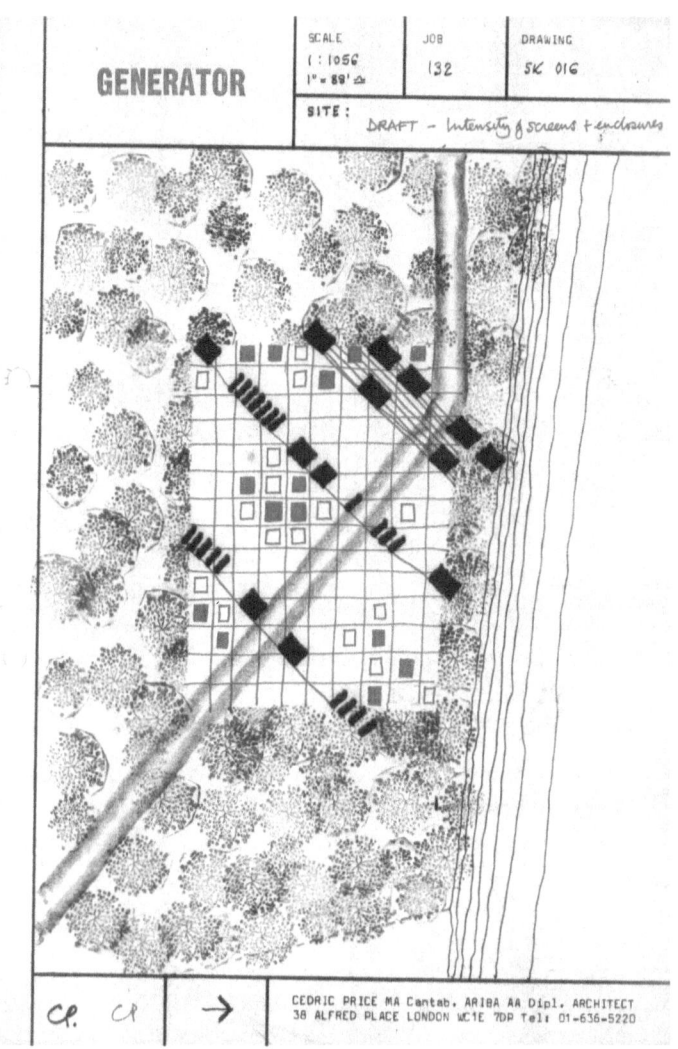

FIGURE 5.4 Plan of Generator project, Cedric Price, 1976–1980. DR1995:0280:190 Cedric Price fonds, Canadian Centre for Architecture © CCA.

and connected to a program. Like in a real-life computer game, the Frazers imagined users organizing different sets and atmospheres with the aid of a simple interface. But in the case that human users were not interesting enough, the building machine would switch on a Boredom Program, in which it would generate unsolicited organizations, learning from past

FIGURE 5.5   Generator project, view of mock-up, unknown photographer,
1979. DR2004:1265:006 Cedric Price fonds, Canadian Centre
for Architecture © CCA.

interactions and gradually improving its accuracy. The Frazers insisted that
this was not merely a spectacle of Intelligence but that the building can
be described as "being literally 'intelligent'" and should "have a mind of
its own".[47]

Price, who once famously posited that "technology is the answer, but
what was the question?", took a more nuanced approach. Uncertain that
it would be so easy to understand how the building operates upon first
contact, he introduced the role of *polariser*, which Jakobson was supposed
to be playing: something between a host and a manipulator, encouraging
people to explore the combinatorial logic of Generator. For him, the

intelligence of Generator would not arise spontaneously, based entirely on its own sophistication, but would have to take its cues from the surrounding natural environment. Boylston Landau, Price's friend and collaborator, would later write that "Generator explores the notion of artificial intelligence, in which the environment itself becomes an intelligent artefact."[48] Indeed, Price's design development process demonstrates an interest in the signals of the surrounding natural system. Early sketches show, among other things, a study of the growth rate and spacing of local slash pine, perhaps anticipating concepts of modules, growth, and flexibility. While the project is shown in most cases as a forest clearing, a rectangle playground amid a dense canopy, many documents almost subconsciously infer that the idea of relentless change comes not only from Generator's programming but from the forest on site. One specifically telling drawing in the project's files shows a section of two forest sites enveloped in amorphous ephemeral shapes and connected via a series of bubbles that seem to represent a communication of sorts. Price's caption reads: "world-wide two way feeds", signaling perhaps a latent ambition to expand the boundaries of the project and its intelligence beyond the

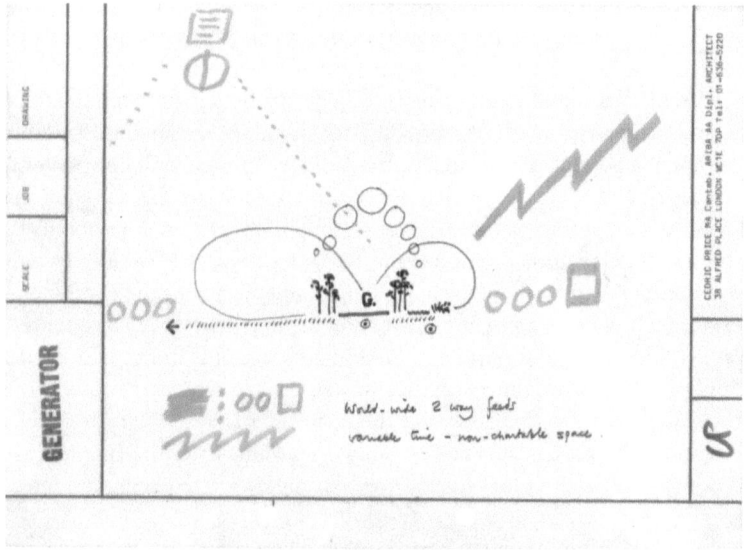

FIGURE 5.6   Sketch for Generator, Cedric Price, 1976–1980. DR1995:0280: 162 Cedric Price fonds, Canadian Centre for Architecture © CCA.

confines of this specific forest site and tap into a global environmental intelligence (Figure 5.6).

The project was abruptly shut down as Gilman was disinherited by his mother and brother. It would take him some years to win the legal battles and revive his dream without Price and the Frazers as a less sophisticated, more conventional version of his piano in the wilderness: a dance center created for his close friend Mikhail Baryshnikov and a wildlife preserve where he would wonder among reticulated giraffes or black rhinos and amuse guests such as Bill Clinton or Isabella Rossellini until his death on-site several years later.[49] The unrealized Generator remains a proposition to consider intelligence as a diffused presence, found in the designed, yet not entirely controlled, dynamic symbiosis between humans, machines, and plants, coevolving in the same forest.

★★★

This diffused or distributed forest intelligence, based as it was on a large number of sensors that transmit information into a central program that generates real-time responses, was largely speculative in the time Generator was terminated. However, a decade later, it became plausible from a technical point of view through the introduction of ubiquitous computing. Mark Weiser, who coined the term in the late 1980s, developed the idea in his work as head of the computer science laboratory at Xerox PARC, where his team attempted to develop new typologies of computers that would vanish into the background. These typologies ranged in size from 1-inch machines to yard-scale displays and were designed to be seamlessly woven into the environments of everyday life. With this "ubiquitous, invisible computing" concept, Weiser sought to present a paradigmatic shift in thinking about computers in the world, one that would not only overcome present technological challenges but answer to the psychological difficulties of living in an information-saturated world.[50] "In essence", he wrote in a text for *Scientific American* that would become a classic, "only when things disappear in this way are we freed to use them without thinking and so to focus beyond them on new goals."[51] The seemingly contradictory proposition of saturating environments with computers in order to make them more natural was explained by referencing the forest:

> There is more information available at our fingertips during a walk in the woods than in any computer system, yet people find a walk among trees relaxing and computers frustrating. Machines that fit the human environment, instead of forcing humans to enter theirs,

will make using a computer as refreshing as taking a walk in the woods.[52]

Rudimentary as this analogy was, it reverberated a view of forests as environments filled with information that constantly flows between trees and humans.

This idea met actual forests in the 1990s, as scientists began applying wireless sensing technologies to natural environments in the form of embedded networked sensing systems. Such a manner of gathering and processing information presented a third way between remote sensing and forest sampling. While technological advances in aerial and satellite photography, largely attributable to the capability to capture and interpret multiple bands of the electromagnetic spectrum, transformed the ability to perform high-fidelity analysis of forest cover, type, and health, these methods remained limited in contexts of multi-storied forests, where the layers beneath the canopy remained unknowable.[53] However, field studies in sample tracts, a common practice since the early days of forestry, were arduous and often inaccurate. While they allowed for sufficient precision in limited areas of experimental forests or tree farms, the practice of extrapolating information gathered from the sample plots onto the entire forest environment in question – now understood as heterogeneous and sensitive to micro-conditions – was inadequate for determining even simple factors such as the number of trees on site. By contrast, embedding low-cost measurement devices in the physical environment and networking them to achieve coordinated sensing allowed for the creation of modular and responsive networks for data flow, accumulation, and analysis, potentially leading to better predictions and new models. Computer scientist Deborah Estrin noted that the processing part of these systems, relying on "multiscale, multimodal, and in-network processing algorithms" was seen not only as optimizing the capacity to answer existing scientific questions but also as opening a realm of new ones that could not be previously asked.[54] At this point, some researchers referred to their emerging integrated sensing systems as intelligent.[55] This was perhaps not surprising: as the devices and architectures of embedded environmental sensing were hypothesized and developed in the hubs of American information technology, they were saturated by the contemporary parlance of smart cities, smart spaces, smart building materials, and smart dust.[56]

One of the research areas on which Estrin focused attention at the Center for Embedded Networked Sensing was Terrestrial Ecology Observing Systems. The major test site for the deployment of these sensing technologies was the experimental forest in James Reserve at the San

Jacinto Mountains in California. Jennifer Gabrys, visiting the site circa 2012, described it as "emerg[ing] through a distribution of sensing processes across organisms, ecological processes, sensing technologies in the form of computational hardware and software, online interfaces, conservation infrastructures, resident scientists, environmental change, citizen scientists, publics, and visiting researchers".[57] Building on posthumanist media and philosophy concepts, Gabrys proposed that the James Reserve forest is indicative of a planetary process of "instrumenting or programming the Earth", one in which "environments, humans and more-than-humans emerge as perceiving and perceivable entities". The sensor-saturated natural environment, part of the smartification program declared in Silicon

FIGURE 5.7   Diagram of James Reserve Experimental Forest, 2008. Illustration by Frank Ippolito.

Valley two decades prior, becomes in this view a transformative setting for human experience and perception (Figure 5.7).[58]

★★★

While networked forests may have been regarded from one perspective as a subset of a wider application program, the idea of forest intelligence received a thrust from ecologists working in the field. When forest scientist Suzanne Simard, during her doctoral work at Oregon State University, used isotope tracers to show that different species of trees in the forest exchange carbon through their interconnected mycelia, the scientific understanding of forests began to catch up with earlier intuitions of naturalists and foresters that spoke of collaboration and exchange between trees. Simard et al. applied techniques used in earlier laboratory experiments but set to prove a set of reciprocal, rather than one-directional, connections between different tree species, through which carbon, nitrogen, or phosphorus flow from "donor" to "receiver" plants.[59] The editors of *Nature*, who published Simard's findings, were tempted to see the hypha connections she followed as equivalent to the webs of communication that undergird contemporary societies. Running the story on the cover of the journal in August 1997, they titled it *The Wood-Wide Web*, exposing a desire to create bridges between biology and technology.[60] The clumsy pun caught on, spreading in scientific journals and popular publications, and gradually finding its way into the language of nongovernmental organizations and public agencies.[61] But this eventual success only came after a period of severe criticism of Simard's findings. The pushback by fellow scientists was based, she wrote, on "the belief that competition was the only plant-to-plant interaction that mattered".[62] Beyond a critique of methodology, the resistance may have had something to do with a deeper nerve the work was touching on. As Simard furthered her research, describing intricate relationships between trees that shared genetic material or the different rates of response by which species exchanged carbon to supply individual trees in need, her work brought up in full force the old ghosts of plant sentience.[63]

Adherents of the idea that plants communicate in ways that can be regarded as intelligent often cite Charles Darwin's late work on plants as an ultimate source of authority (Figure 5.8). Following a series of innovative experiments on root systems and patterns of plant growth with his son Francis, published as a book in 1880, Darwin concluded:

> It is hardly an exaggeration … to say that the tip of the radicle … having the power of directing the movements of the adjoining parts,

acts like the brain of one of the lower animals; the brain being seated within the anterior end of the body, receiving impressions from the sense organs and directing the several movements.[64]

The plant brain hypothesis, coming from the most influential scientist of the era, and blurring established boundaries between the animal and plant kingdoms was indeed striking. But it also rehearsed an ages-old heated scientific and philosophical debate about precisely these boundaries. A well-known point in this conversation is found in Aristotle's treatise on plants, in which it is written: "I assert, then, that plants have neither sensation nor desire."[65] The decisiveness of this statement, much like the ones being made in the twenty-first century, was geared towards the rebuttal of contemporary thinkers who held different opinions. The treatise details what the author regards as the folly in assigning plants intellect and intelligence, betraying that the matter was far from settled.[66] What is also evident in the treatise is the fact that the writer's argumentation is much superior to his observations, as he bases his rejection of plant sentience on its inability to sense and communicate. The Aristotelian line supposedly inspired a history of clear delineation between plant and animal that stratified a Western scientific tradition as opposed to other ways of thinking. However, contrary to common conceptions that draw a clear boundary between East and West regarding the spiritual and cultural significance assigned to all living things, Eastern traditions often display ambivalence toward the question of sentient plants. It is true that Jainism sometimes considers plants as living beings, and thus subject to the ethic of nonviolence, and that in the Hindu Mahābhārata, trees are mentioned as possessing the five senses, being "susceptible of pleasure and pain".[67] However, Buddhist writings of the Pāli Canon are "indefinite at best" and generally exclude plants from its categorization of living things.[68] It took almost a millennium for plants and trees to begin to be considered as sentient by East Asian Buddhist thinkers in China and later in Japan.[69] Different schools of thought debated the issue for many decades before the idea of including plants within the realm of those with the potential of Buddhahood became more accepted.[70] Such ideas, assimilating earlier forms of *kami* worship in indigenous Japanese religions, developed into cultural and spiritual notions of kinship between humans and plants. Eventually, this led to the far-reaching proposition that plants could be regarded as models for man in his quest for enlightenment and Buddhist realization.[71] Other indigenous spiritual traditions demonstrate a similar ambiguity. Simard mentions a belief, prevalent among the Coast Salish peoples of the Pacific Northwest, that trees have personhood and that the forest is made of many interconnected nations, which inspired her own

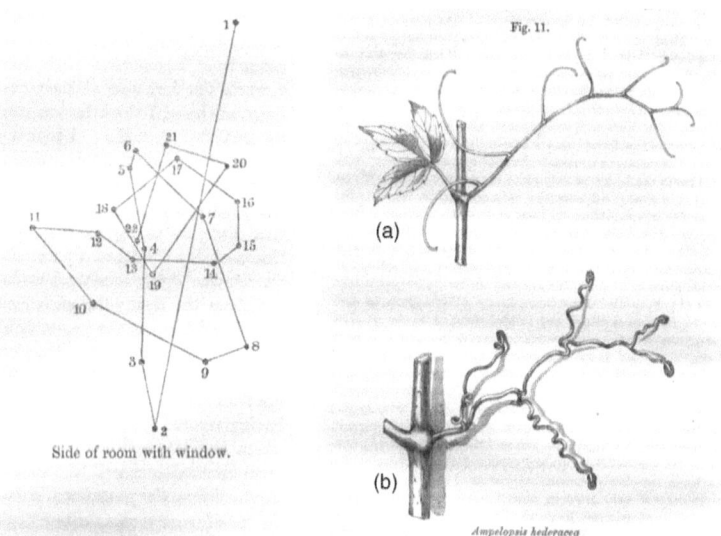

FIGURE 5.8   Plant movement diagram, from Charles and Francis Darwin's *The Power of Movement in Plants*, 1890.

thinking. However, the reference to the ontology of all Native American cultures as encapsulating a different understanding of plants is far from grounded. For instance, while plant usage for medicinal and spiritual purposes in pre-Columbian cultures is estimated to date several millennia, their status as fellow beings often escapes a clear definition, varying widely between different groups and times.[72]

The science of individual plant communication continues to develop, resulting in researchers that openly speak of sentient plants.[73] But it was these previously mentioned extrascientific ways of knowing that became important to the specific metaphor of forest systems as sites of diffused intelligence. Anthropologists peering into the life of indigenous communities buttressed this idea when proposing that the peoples they were observing were interconnected with flora and fauna in inexplicable ways. When Eduardo Kohn ventured in the 1980s to live among the Runa people of the upper Amazon, he was taken by the ways in which they navigate and make sense of their environment. Immersing himself in the field, Kohn gathered revealing insights that rendered plausible the notions of a collective forest mind: the Runa were communicating with their dogs, meandering through dreams and reality, transforming into other beings of the forest. Gradually, he was moved to think that signals flowed between animal, vegetal, and human members of these communities without the need for a mycelial network. "The rainforest", he wrote, "comes to be what it is: an emergent and expanding multilayered cacophonous web of mutually constitutive, living, and growing thoughts."[74] In his account, Kohn ventures on an ambitious theoretical attempt to describe an "anthropology beyond the human" – one that can account for the interconnectedness of humans and nonhumans that he witnessed first-hand. To do so, he defines an "ecology of selves" – "the web of living thoughts in and around the forests … which includes not only the many kinds of living beings of the forest but also … the spirits and the dead that make us the living beings that we are."[75] These provocative assertions, novel as they sound, are in fact building on Western semiotic frameworks. Kohn uses readings of Charles Sanders Pierce's early twentieth-century theory of signs to claim that "all life is semiotic, and all semiosis is alive". From that follows that all living selves, "bacterial, floral, fungal, or animal" are equal in that they represent the world around them in some way which is constitutive of their being.[76] In that, Kohn follows the footsteps of anthropologists such as Margaret Mead or Thomas Sebeok, who in the early 1960s adapted and popularized Pierce's ideas as "semiotics", researching "patterned communication in all modalities". These included, in Seboek's case, signaling behaviors in and across animal species.[77] The claim that forests *really* think situated Kohn's work in a contemporary discourse of the posthumanities and attracted both praise and critique.[78] However, while it is imbued with many of the tropes found in that discourse, Kohn's depictions of the tropical forest, and specifically his application of concepts proposed by Gregory Bateson, bear a strange resemblance to the idea of information-dense environments that inspired the design of projects such as Generator or the IBM pavilion. The lingering question connecting these projects is where exactly intelligence is to be found and at what point does it become a forest mind.

★★★

In the white paper describing Terra0, the authors took an explicitly technical stance, avoiding any implications of intelligence when describing the self-monitoring, self-replicating forest, and focusing on economic mechanisms – the smart contract that negotiates financial transactions, the initial crowdsale phase in which the initiators invest Ether in exchange to Terra0 tokens, and the ways through which the amount of wood to be sold can be automatically estimated. From this perspective, the entire enterprise can be understood as simply carrying forward two hundred years of forest abstraction and calculation to its logical conclusion. But there is a second, more implicit, assumption that undergirds the blockchain utopia imagined by the authors, and that is that their smart forest, now freed from human intervention, will be able to make the best decisions for its own survival and future growth. In this, it shares the intuition that humanity's poor track record in managing its own and other species' survival could only be corrected by nonhuman intelligence, present at least since Fuller's World Brain. As threatening as a world dominated by artificial, omnipresent intelligence may be, Terra0 could perhaps offer some form of solace.[79] As if against the will of its human operators, the project's website was filled with excerpts from a book by American poet Richard Brautigan, which portray a romantic fusing of organic and nonorganic life: "I would like to think of a cybernetic forest ... filled with pines and electronics where deer stroll peacefully past computers as if they were flowers with spinning blossoms."[80] This depiction, uncannily resembling earlier ideas of Buddhist enlightenment and later visions of ubiquitous computing, lead to salvation as Brautigan writes: "I like to think of a cybernetic ecology, where we are free of our labors and joined back to nature, returned to our mammal brothers and sisters, and all watched over by machines of loving grace."[81] Diffused forest intelligence, a complex metaphor emerging at the meeting points of ecology, cybernetics, anthropology, and spatial design, could, in this view, not only sustain an environment or alter a perception but also work its mysterious ways to undo centuries of folly and avarice and reunite us all with lost Nature.

Notes

1 Seidler, Paul, Paul Kolling, and Max Hampshire. "terra0: Can an Augmented Forest Own and Utilise itself?" Berlin University of the Arts, 2016. https://terra0.org/assets/pdf/terra0_white_paper_2016.pdf. Accessed March 28, 2024.

2 *Te Awa Tupua (Whanganui River Claims Settlement) Act 2017*, Public Act, 2017, No. 7, assented to March 20, 2017. (New Zealand).

3 O'Donnell, Erin L., and Julia Talbot-Jones. "Creating Legal Rights for Rivers: Lessons from Australia, New Zealand, and India." *Ecology and Society* 23, no. 1 (2018). https://www.jstor.org/stable/26799037.

4 In some respects, the declarative value of these acts is as important to their ameliorative effects as their implementation. However, granting legal rights to natural entities does not always address the full spectrum of historical land access and ownership issues. See Collins, Toni, and Shea Esterling. "Fluid Personality: Indigenous Rights and the Te Awa Tupua (Whanganui River Claims Settlement) Act 2017 in Aotearoa New Zealand." *Melbourne Journal of International Law* 20, no. 1 (July 2019): 17.

5 Stone, Christopher D. "Should Trees Have Standing? – Towards Legal Rights for Natural Objects." *Southern California Law Review* 45 (1972): 453.

6 Here, too, Stone makes a striking comparison between the current status of rights natural elements have in court and similar discussions regarding enslaved Black people in American courts, which are foreign to our current sensibilities.

Stone, Christopher D. "Should Trees Have Standing? – Towards Legal Rights for Natural Objects." *Southern California Law Review* 45 (1972): 455.

7 "I am proposing we do the same with eagles and wilderness areas as we do with copyrighted works, patented inventions, and privacy: make the violation of rights in them to be a cost by declaring the 'pirating' of them to be the invasion of a property interest. If we do so, the net social costs the polluter would be confronted with would include not only the extended homocentric costs of his pollution (explained above) but also costs to the environment per se" Stone, Christopher D. "Should Trees Have Standing? – Towards Legal Rights for Natural Objects." *Southern California Law Review* 45 (1972): 476.

8 Stone, "Should Trees Have Standing?", 500.

9 Pinchot, Gifford. *A Primer of Forestry: The Forest.* Washington: Govt. Print Office, 1899, 43.

10 Pinchot, Gifford. *A Primer of Forestry: The Forest.* Washington: Govt. Print Office, 1899, 43.

11 Pinchot, Gifford. *A Primer of Forestry: The Forest.* Washington: Govt. Print Office, 1899, 45.

12 Pinchot, Gifford. *A Primer of Forestry: The Forest.* Washington: Govt. Print Office, 1899, 47. For more on these undercurrents in Pinchot's *Primer*, see Handel, Dan. "The Forest Community: Sovereign, Subject, Trees." *Harvard Design Magazine* 45 (Spring/Summer 2018): 202–206.

13 Leopold writes, "All ethics … rest upon a single premise: that the individual is a member of a community of interdependent parts. His instincts prompt him to compete for his place in the community, but his ethics prompt him also to co-operate." The land ethic is a projection of that idea onto the nonhuman world, "simply enlarg[ing] the boundaries of the community to include soils, waters, plants, and animals, or collectively: the land." Leopold, Aldo. "The Land Ethic." In *A Sand County Almanac*, 203–204. New York: Oxford University Press, 1989 [1949].

14 Sierra Club v. Morton, 405 U.S. 727 (1972), 752.

15 Sierra Club v. Morton, 405 U.S. 727 (1972), 743.

16 While the United States government introduced environmental laws in the early 1970s, its courts lagged behind those of Bolivia or Ecuador in securing the rights of the environment. See Daly, Erin. "The Ecuadorian Exemplar: The First Ever Vindications of Constitutional Rights of Nature." *Review of European Community & International Environmental Law* 21, no. 1 (April 2012): 66.

17 In his dense survey of the state of the art of deep ecology circa 1987, George Sessions writes:

A number of developments have emerged as 'cutting edge' issues in ecophilosophy that can roughly be described as environmental ethics versus an ecological metaphysics, world view, or ontology of being; ethical hierarchies versus an egalitarian position; animal rights ethical theorizing versus an overall ecological world view; the rights of ecofeminism; the rise of the 'new physics' and its relation to ecophilosophy; and the difference between the New Age/Aquarian Conspiracy and Deep Ecology. As will be evident, these issues overlap to a considerable degree.

Sessions, George. "The Deep Ecology Movement: A Review." Environmental Review: ER 11, no. 2 (1987): 115

18 These attempts often followed mathematician and game theorist John Von Neumann, who sketched possible scenarios of Allies' victory in the 1940s and, after his wartime work on the Manhattan Project, went on to develop the digital computer.

19 Fuller's focus on strategic games and scenario building was part of a contemporary scene of complex multi-actor simulations, famously manifest in the Cold War games of the RAND Corporation played in the mid-1950s. See Ferng, Jennifer. "Designing Conclusions for a Cold War Humanity." *Thresholds*, no. 32 (2006): 67.

20 Buckminster Fuller Institute. "World Game." Accessed April 4, 2024. https://www.bfi.org/about-fuller/big-ideas/world-game/.

21 "Fuller himself admitted that the finalized version of the World Game would never be played until the simulations themselves could be calculated on a computer." Ferng, "Designing Conclusions," 68.

22 Fuller's student and colleague Medard Gabel developed in the 1970s several game projects using optimistic scenario planning. Ferng, "Designing Conclusions," 69.

23 Fuller, R. Buckminster. "The World Game." *Ekistics* 28, no. 167 (1969): 291.

24 Turing, Alan M. "Computing Machinery and Intelligence." *Mind* 59, no. 236 (1950): 433.

25 These include the ELIZA and PARRY computer programs, created in 1966 and 1972, respectively, as experiments in artificial psychotherapy, which are the forerunners of much of the current landscape of chatbots.

26 In describing the influences of cybernetic ideas on architect Christopher Alexander, musician Brian Eno, and writer William Burroughs, among others, Andrew Pickering writes: "Sociologically, then, cybernetics wandered around as it evolved, and I should emphasize that an undisciplined wandering of its subject matter was a corollary of that." Pickering, Andrew. *The Cybernetic Brain: Sketches of Another Future.* Chicago: University of Chicago Press, 2010, 11.

27 Licklider was not to first to hypothesize a future in which machine assists and extends human thinking. Scientist Vannevar Bush imagined in 1945 a device called *memex* on which an individual can store large amounts of information and create personalized "association trails". However, this machine remained auxiliary to human thinking processes. See Bush, Vannevar. "As We May Think," *Atlantic Monthly* 176 (July 1945): 101–108.

28 Licklider writes that the work will be generally allocated to the parties: "men will set the goals, formulate the hypotheses, determine the criteria, and perform the evaluations. Computing machines will do the routinizable work that must be done to prepare the way for insights and decisions in technical and scientific thinking." Licklider, J. C. R. "Man-Computer Symbiosis." *IRE*

Transactions on Human Factors in Electronics HFE-1 (March 1960): 4.

29 Licklider, J. C. R. "Man-Computer Symbiosis." *IRE Transactions on Human Factors in Electronics* HFE-1 (March 1960): 4.

30 "Architects Praise 4 Pavilions at Fair." *New York Times*, November 12, 1964, 61. "The People Wall whisks you up into the giant egg, where the Information Machine reveals that you too can be a computer, of sorts." "The New York Fair." *Time*, August 14, 1964, Vol. 84, no. 7. Accessed April 7, 2024. https://content.time.com/time/subscriber/article/0,33009,897221,00.html.

31 Ada Louise Huxtable. "Architecture: Chaos of Good, Bad and Joyful." *New York Times*, April 22, 1964, 25.

32 "The Thinking Man's Exhibit." *Esquire*, October 1963, 118.

33 Colomina, Beatriz. "Enclosed by Images: The Eameses' Multimedia Architecture." *Grey Room*, no. 2, The MIT Press, 2001, 20.

34 See Roche, Kevin. "Charles Eames." In *Zodiac*, no. 11 (March 1994): 114–21.

35 At the time, the investment in the 360 system was considered a "bet-the-business" move.

36 Charles and Ray Eames. *The IBM Pavilion*. Armonk, New York: International Business Machines Corporation, 1964, 3.

37 Quoted in John Harwood's illuminating essay on the instrumental role of Noyes in the definition of information architecture. See Harwood, John. "The White Room: Eliot Noyes and the Logic of the Information Age Interior." *Grey Room*, no. 12 (Summer 2003): 13.

38 Banham, Reyner. *The Aspen Papers; Twenty Years of Design Theory from the International Design Conference in Aspen*. New York: Praeger, 1974, 11.

39 This understanding was close to the one expressed by Noyes and mentioned earlier. For a description of the changing meanings of the term within and around design, see Martin, Reinhold. "Environment, c. 1973." *Grey Room*, no. 14 (Winter 2004): 83.

40 For a vivid account of the chaotic unfolding of the 1970 conference, see Twemlow, Alice. "A Guaranteed Communications Failure: Consensus Meets Conflict at the International Design Conference in Aspen, 1970." In *Aspen Complex*, edited by Martin Beck, 110–37. Berlin: Sternberg Press, 2013.

41 Banham, *The Aspen Papers*, 207.

42 Reinhold Martin notes that this presented an updating of Marx's ideas regarding Nature as expressed in his earlier, well-known book *The Machine in the Garden*, softening the opposition between nature and civilization and integrating humans in the ecosystem. See Martin, "Environment c. 1973," 82.

43 The collection, containing more than two hundred drawings including some of Price's, was gifted to the Museum of Modern Art soon after Gilman's passing.

44 The brief was written down on a napkin during a plane ride back from the site. See Antonelli, Paola. "Interview with Pierre Apraxine." In *The Changing of the Avant-Garde: Visionary Architectural Drawings from the Howard Gilman Collection*, edited by Terence Riley, 150. New York: Museum of Modern Art, 2002.

45 Steenson, Molly Wright. 2017. *Architectural Intelligence: How Designers and Architects Created the Digital Landscape*. Cambridge, Mass.: The MIT Press, 128.

46 A series of Polaroid photos in the Cenric Price Fonds at the Candian Centre for Architecture show the wood mock-ups and a crane that was brought it for the occasion to simulate something of what the project could be like against the forest setting.

47 Steenson, *Architectural Intelligence*, 147.

48 Landau, Royston. "An Architecture of Enabling – The Work of Cedric Price." *AA Files* 8 (Spring 1985): 7.

49 Lenzner, Robert, and Tomas Kellner. "The Fall of the House of Gilman." August 11, 2003. *Forbes*. Accessed April 9, 2024. https://www.forbes.com/forbes/2003/0811/068.

50 Weiser describes the required technology for the application of his concept as consisting of three parts: "cheap, low-power computers that include equally convenient displays, a network that ties them all together, and software systems implementing ubiquitous applications". See Weiser, Mark. "The Computer for the 21st Century." *Scientific American* 265, no. 3 (1991): 100.

51 Weiser, Mark. "The Computer for the 21st Century." *Scientific American* 265, no. 3 (1991): 94.

52 Weiser, Mark. "The Computer for the 21st Century." *Scientific American* 265, no. 3 (1991): 104.

53 Remote sensing emerged from aerial military reconnaissance, which involved systematic photogrammetry and photo interpretation. By the early 1960s, the technologies and expertise involved matured to be considered for applications in fields such as urban planning, forestry, hydrology, wildlife management, and tax assessment. The multispectral abilities of new cameras were especially useful in forest areas as loss of infrared reflectance could indicate blight by damaging agents before it could be seen on site. For an overview of the state of research and industry circa 1964, see Colwell, Robert N. "Aerial Photography – A Valuable Sensor for the Scientist." *American Scientist* 52, no. 1 (1964): 16–49. By the end of the decade, many of the people and institutions working on remote sensing coalesced around the Earth Resources Technology Satellite program, later known as Landsat.

54 Estrin, a computer scientist specializing in network architecture, turned her attention circa 2000 to embedded sensory networks and co-founded in 2002 the National Science Foundation (NSF)–funded Center for Embedded Networked Sensing at UCLA. See Estrin, Deborah. "Reflections on Wireless Sensing Systems: From Ecosystems to Human Systems." UCLA: Center for Embedded Network Sensing, 2007. Accessed April 12, 2024. https://escholarship.org/uc/item/77r2j6p0.

55 Estrin, Deborah. "Reflections on Wireless Sensing Systems: From Ecosystems to Human Systems." UCLA: Center for Embedded Network Sensing, 2007. Accessed April 12, 2024. https://escholarship.org/uc/item/77r2j6p0.

56 Estrin, Deborah, Ramesh Govindan, and John Heidemann. "Embedding the Internet." *Communications of the ACM* 43, no. 5 (May 2000): 39–41.

57 Gabrys, Jennifer. "Sensing an Experimental Forest: Processing Environments and Distributing Relations." *Computational Culture* 2 (September 28, 2012). http://computationalculture.net/sensing-an-experimental-forest-processing-environments-and-distributing-relations/.

58 There is a fine line between sensing and knowing, the crossing of which opens a thread of describing forests as "juridicial courts" or "technical laboratory". See Biemann, Ursula, and Paulo Tavares. *Forest Law = Selva Jurídica*. Michigan: Eli and Edythe Broad Art Museum, 2014, 58. Brought up and discussed in Jacobs, Daniel, and Brittany Utting. "Forest Governmentality and the Struggle for More-Than-Human Sovereignty." In *The Avery Review* 56 (April 2022). http://averyreview. com/issues/56/forest-governmentality-and-the-struggle-for-more-than-human-sovereignty.

59 Simard, Suzanne, David Perry, Myron Jones, et al. "Net Transfer of Carbon Between Ectomycorrhizal Tree Species in the Field." *Nature* 388 (1997): 579.

60 The term did not appear in the original article.

61 Scientific texts adopted almost immediately the term when referring to mycorrhizal hyphae in forests. Two decades later, it is possible to read on the One Earth organization website that "beneath every forest lies a web of millions of organisms passing information"; listen to a BBC narrator explaining that "trees are secretly talking, trading, or waging war on one another"; and learn that "plants use an internet made of fungus" on PBS.

62 The *Nature* experiment was eventually redone by fellow scientists to answer to the critics. See Simard, Suzanne. *Finding the Mother Tree: Discovering the Wisdom of the Forest*. 1st ed. New York: Alfred A. Knopf., 2021, ch. 11, Kindle.

63 In later years, and outside of the confines of scientific publications, Simard would openly speak and write about forest systems in which Mother Trees act as "the majestic hubs at the center of forest communication, protection, and sentience". Simard, Suzanne. *Finding the Mother Tree: Discovering the Wisdom of the Forest*. 1st ed. New York: Alfred A. Knopf., 2021, Introduction.

64 Darwin, Francis, and Charles Darwin. *The Power of Movement in Plants*. London: John Murray, 1880, 573.

65 It is debatable whether Aristotle wrote the version that we now have. See Barnes, Jonathan. *The Complete Works of Aristotle*. Vol. 2. Princeton, NJ: Princeton University Press, 1984, 1252.

66 The writer focuses on pre-Socratic thinkers, especially Anaxagoras, Democritus, and Empedocles, who each attributed various forms of consciousness to trees and other plants.

67 Ganguli, Kisari Mohan, trans. *The Mahābhārata*. Calcutta: Bharata Press, 1884, 12.184.

68 Findly, Ellison Banks. "Borderline Beings: Plant Possibilities in Early Buddhism." *Journal of the American Oriental Society* 122, no. 2 (2002): 253.

69 The discussions around plants begin with the Chinese hypothesis regarding a universalist Buddha nature of all things and change as it moves to Japan into ideas about a special connection with natural elements. See La Fleur, William R. "Saigyō and the Buddhist Value of Nature. Part I." *History of Religions* 13, no. 2 (1973): 95.

70 Significantly, at this point in tenth-century Japan, rocks and rivers were excluded from this realm. La Fleur, William R. "Saigyō and the Buddhist Value of Nature. Part I." *History of Religions* 13, no. 2 (1973): 104.

71 LaFleur, William R. "Saigyō and the Buddhist Value of Nature. Part II." *History of Religions* 13, no. 3 (1974): 227.

72 The growing body of literature on hallucinogenic substances in Mesoamerican cultures, prompted by a growing commercial and scholarly interest in their applications in Western societies, is often tainted by cultural bias. This is part of the inherent difficulty of translating ingenious cosmologies into "objective" research terms. In 2016, Joaquin Mana of the Huni Kuin elucidated the point:

> "it is not necessary to explain everything about our knowledge. That can be our own internal knowledge. There are many names for the plant – it is a multicultural matter. Each Indigenous people has their own names and stories. Researchers sometimes generalize all of this diverse knowledge."
>
> *Quoted in Dev, Laura. "Plant Knowledges: Indigenous Approaches and Interspecies Listening Toward Decolonizing Ayahuasca Research." In Plant Medicines, Healing and Psychedelic Science: Cultural Perspectives, edited by Beatriz Caiuby Labate and Clancy Cavnar. Cham: Springer International Publishing, 2018, 195*

73 These developments led to the emergence of plant neurobiology, a field that attempts to explain the complex behaviors of intelligent plants. See the oft-quoted, yet highly controversial, article: Brenner, Eric D., Rainer Stahlberg, Stefano Mancuso, Jorge Vivanco, František Baluška, and Elizabeth Van Volkenburgh. "Plant Neurobiology: An Integrated View of Plant Signaling." *Trends in Plant Science* 11, no. 8 (2006): 413–9.

74 Kohn, Eduardo. *How Forests Think: Toward an Anthropology Beyond the Human.* Berkeley: University of California Press, 2013, 79.

75 Kohn, Eduardo. *How Forests Think: Toward an Anthropology Beyond the Human.* Berkeley: University of California Press, 2013, 78.

76 Kohn, Eduardo. *How Forests Think: Toward an Anthropology Beyond the Human.* Berkeley: University of California Press, 2013, 6.

77 Mead proposed the term semiotics following a 1962 conference in Indiana University. Seboek was working at the same time on developing zoosemiotics. Sebeok, Thomas A. "Zoosemiotics." *American Speech* 43, no. 2 (1968): 142.

78 The book makes sure to cite many of the central figures in the posthumanist turn, including Bruno Latour, Giorgio Agamben, and Donna Haraway.

79 Paul Seidler of Terra0, using a cultural reference from the apocalyptic block-buster The Terminator, explained in one interview that the project partly came from his need to "save himself when Skynet finally arrives". Raponi, Martina. "Terra0, The Augmented Self-Owned Forest." March 14, 2017. https://digicult.it/news/terra0-la-foresta-aumentata-independente/. Accessed April 16, 2024.

80 Terra0. https://terra0.org/. Accessed April 16, 2024.

81 The quotes are taken from Brautigan's 1967 poem, published in a collection under the same name, and acquiring a cult status among cybernetics and data enthusiasts.

INDEX

Pages in *italics* refer to figures, and pages followed by n refer to notes.